REPORT TO THE PRESIDENT
ACCELERATING U.S. ADVANCED MANUFACTURING

Executive Office of the President

President's Council of Advisors on Science and Technology

October 2014

About the President's Council of Advisors on Science and Technology

The President's Council of Advisors on Science and Technology (PCAST) is an advisory group of the Nation's leading scientists and engineers, appointed by the President to augment the science and technology advice available to him from inside the White House and from cabinet departments and other Federal agencies. PCAST is consulted about, and often makes policy recommendations concerning, the full range of issues where understandings from the domains of science, technology, and innovation bear potentially on the policy choices before the President.

For more information about PCAST, see www.whitehouse.gov/ostp/pcast

The President's Council of Advisors on Science and Technology

Co-Chairs

John P. Holdren
Assistant to the President for
Science and Technology
Director, Office of Science and Technology
Policy

Eric Lander
President
Broad Institute of Harvard and MIT

Vice Chairs

William Press
Raymer Professor in Computer Science and
Integrative Biology
University of Texas at Austin

Maxine Savitz
Vice President
National Academy of Engineering

Members

Rosina Bierbaum
Dean, School of Natural Resources and
Environment
University of Michigan

S. James Gates, Jr.
John S. Toll Professor of Physics
Director, Center for String and Particle
Theory
University of Maryland, College Park

Christine Cassel
President and CEO
National Quality Forum

Mark Gorenberg
Managing Member
Zetta Venture Partners

Christopher Chyba
Professor, Astrophysical Sciences and
International Affairs
Director, Program on Science and Global
Security
Princeton University

Susan L. Graham
Pehong Chen Distinguished Professor
Emerita in Electrical Engineering and
Computer Science
University of California, Berkeley

Shirley Ann Jackson
(*through September 2014*)
President
Rensselaer Polytechnic Institute

Michael McQuade
Senior Vice President for Science and Technology
United Technologies Corporation

Chad Mirkin
George B. Rathmann Professor of Chemistry
Director, International Institute for Nanotechnology
Northwestern University

Mario Molina
Distinguished Professor, Chemistry and Biochemistry
University of California, San Diego
Professor, Center for Atmospheric Sciences at the Scripps Institution of Oceanography

Craig Mundie
Senior Advisor to the CEO
Microsoft Corporation

Ed Penhoet
Director, Alta Partners
Professor Emeritus, Biochemistry and Public Health
University of California, Berkeley

Barbara Schaal
Mary-Dell Chilton Distinguished Professor of Biology
Washington University, St. Louis

Eric Schmidt
Executive Chairman
Google, Inc.

Daniel Schrag
Sturgis Hooper Professor of Geology
Professor, Environmental Science and Engineering
Director, Harvard University Center for Environment
Harvard University

Staff

Marjory S. Blumenthal
Executive Director

Knatokie Ford
AAAS Science & Technology Policy Fellow

Ashley Predith
Assistant Executive Director

The President's Council of Advisors on Science and Technology

Accelerating U.S. Advanced Manufacturing

AMP2.0 Steering Committee Report

★ ★

★ ★

EXECUTIVE OFFICE OF THE PRESIDENT
PRESIDENT'S COUNCIL OF ADVISORS ON SCIENCE AND TECHNOLOGY
WASHINGTON, D.C. 20502

President Barack Obama
The White House
Washington, DC 20502

Dear Mr. President,

This letter transmits a report entitled *Accelerating U.S. Advanced Manufacturing*, prepared by the Steering Committee of the Advanced Manufacturing Partnership 2.0 (AMP2.0). The President's Council of Advisors on Science and Technology (PCAST) has reviewed and adopted the report, which follows up the first AMP report, *Capturing Domestic Competitive Advantage in Advanced Manufacturing (July 2012)*, and the 2011 PCAST report, *Ensuring American Leadership in Advanced Manufacturing*.

The members of AMP2.0 worked with industry, academia, labor, government, and the public to address the challenge of expanding advanced manufacturing across the United States. Led by co-chairs Rafael Reif, president of the Massachusetts Institute of Technology, and Andrew Liveris, CEO of the Dow Chemical Company, AMP2.0 built its activities and recommendations on the three pillars established in the 2012 report: (1) enabling innovation, (2) securing the talent pipeline, and (3) improving the business climate.

In the past year, teams of experts assembled by AMP2.0 identified manufacturing technology areas where the United States could establish a strategic advantage. A coalition of community colleges and companies developed a best-in-class apprenticeship model and launched a trial program with colleges in northern California and southern Texas. The Printed Electronics Pilot Project uncovered the types of technical and market information that small- and medium-sized manufacturers need to develop successful scale-up business plans. Five AMP2.0 regional meetings throughout the year and Manufacturing Day on October 3rd drove further momentum across the country.

The project identified a number of further steps the Federal government can take to further U.S. advanced manufacturing capabilities. With the Manufacturing Innovation Institutes as a cornerstone of the Nation's investment, implementing a Federal strategic plan in advanced manufacturing across all Federal activities from the Institutes to individual agency program areas is one important step. Two others are (1) ensuring that advanced manufacturing research addresses questions along the pipeline of technology maturity and (2) leveraging Federal organizations to improve information flow to manufacturers.

To ensure a cohesive Federal effort, PCAST recommends that the Executive Office of the President develop and release, within sixty days, a plan for the implementation of the AMP2.0 report's recommendations.

Advanced manufacturing is a domain of great potential, the achievement of which will require drawing on resources from the public, academic, and industrial sectors all across the country. The Federal government can and should continue to catalyze the needed effort. Thank you for the opportunity to provide input on this important topic.

Best regards,

John P. Holdren
Co-chair, PCAST

Eric S. Lander
Co-chair, PCAST

AMP2.0 Working Group

Rafael Reif
President
Massachusetts Institute of Technology

Andrew Liveris
President, Chairman and CEO
The Dow Chemical Company

Shirley Ann Jackson
Member, PCAST
President
Rensselaer Polytechnic Institute

PCAST Staff

Marjory S. Blumenthal
Executive Director, PCAST

Ashley Predith
Assistant Executive Director, PCAST

National Economic Council Staff

Jason Miller
Special Assistant to the President for Manufacturing

J. J. Raynor
Senior Policy Advisor

Advanced Manufacturing National Program Office (AMNPO) Staff

Michael F. Molnar
Director, AMNPO

Gloria J. Wiens
Assistant Director for Research Partnerships, AMNPO

Advanced Manufacturing Partnership 2.0 Steering Committee

Co-Chairs

Rafael Reif
President
Massachusetts Institute of Technology

Andrew Liveris
President, Chairman and CEO
The Dow Chemical Company

Members

Wes Bush
Chairman, CEO and President, Northrop Grumman

David Cote
Chairman and CEO, Honeywell

Nicholas Dirks
Chancellor, University of California, Berkeley

Kenneth Ender
President, Harper College

Leo Gerard
International President,
United Steelworkers

Shirley Ann Jackson
President, Rensselaer Polytechnic Institute

Eric Kelly
President and CEO, Overland Storage, INC

Klaus Kleinfeld
Chairman and CEO, Alcoa, INC

Ajit Manocha
Senior Advisor, GLOBALFOUNDRIES

Douglas Oberhelman
Chairman and CEO, Caterpillar, INC

Annette Parker
President, South Central College

G.P. "Bud" Peterson
President, Georgia Institute of Technology

Luis Proenza
President Emeritus, The University of Akron

Mark Schlissel
President, University of Michigan

Eric Spiegel
President and CEO, Siemens Corporation

Mike Splinter
Executive Chairman of the Board, Applied Materials, INC

Christie Wong Barrett
CEO, Mac Arthur Corporation

Table of Contents

Executive Summary 1
    ENABLING INNOVATION 2
    SECURING THE TALENT PIPELINE 7
    IMPROVING THE BUSINESS CLIMATE 10
    IMPLEMENTATION 12

I. Advanced Manufacturing Partnership 2.0 13
    Introduction 13
    Process 14
    AMP 2.0 Steering Committee Actions & Recommendations 15
    Summary of the Advanced Manufacturing Partnership 2.0's
    Recommendations 17

II. Recommendations 20
    Pillar 1: Enabling Innovation 20
        Establishing a National Manufacturing Technology Strategy 21
        Establishing & Supporting the National Network for Manufacturing
        Innovation Institutes 29
    Pillar 2: Securing the Talent Pipeline 30
        Shifting the Public's Misconceptions of Manufacturing 31
        Connecting More Americans with Skills for Successful Careers in
        Manufacturing 32
    Pillar 3: Improving the Business Climate 38
        Expanding and Exchanging Intermediary Solutions for Manufacturing 39
        Increasing Capital Access For Established and Start-up Firms 41

III. Conclusions 44

Appendix A: AMP Recommendations and Implementation Status 45

Appendix B: AMP2.0 Membership and Participation 52

Appendix C: U.S. Regional Meetings on Advanced Manufacturing during the
Advanced Manufacturing Partnership 2.0 56

Appendix 1: Transformative Manufacturing Technologies 58

Appendix 2: Demand-Driven Workforce Development and Training 71

Appendix 3: National Network for Manufacturing Innovation (NNMI) Analysis 74

Appendix 4: Scale-Up Policy.. 77

Appendix 5: Manufacturing and Engagement.. 88

Appendix 6: Abbreviation Glossary... 92

Annexes 1-10: Manufacturing Technology Areas..available online

Annexes 11-24: Workforce ....available online

Annexes 25-31: NNMI Analysis..available online

Annex 32: Scale-Up Policy ....available online

Executive Summary

The United States has been the leading producer of manufactured goods for more than 100 years, and the manufacturing sector is once again adding jobs and opening new factories at its fastest rate in two decades. The United States has long thrived as a result of its ability to manufacture goods and sell them to global markets. Manufacturing drives knowledge production and innovation in the United States by supporting two-thirds of private sector research and development and by employing the vast majority of U.S. scientists, engineers, and technicians to invent and produce new products. Yet, in the 2000's, manufacturing faced major employment declines as factories were shuttered. U.S. strengths in manufacturing innovation and technologies that have sustained American leadership in manufacturing are under threat from new and growing competition abroad.

In its July 2012 inaugural report, *Report to the President on Capturing Domestic Competitive Advantage in Advanced Manufacturing*, the first Advanced Manufacturing Partnership (AMP) called for a full court press to increase U.S. competitiveness for advanced manufacturing by sustaining U.S. investments in science, technology, and innovation; establishing a National Network of Manufacturing Innovation Institutes—a set of public-private partnerships to build shared high-tech facilities and advance U.S. leadership in emerging technologies; upgrading community-college workforce training programs and deploying the talent of returning veterans to meet critical manufacturing skills needs; and improving the business climate for manufacturing investment through tax, regulatory, energy, and other policies.

Building upon its initial findings and growing interest in a resurgent U.S. manufacturing sector, the Advanced Manufacturing Partnership was re-chartered (AMP2.0) and has worked with the federal government to implement the highest priority recommendations from its original report. These AMP2.0 efforts from September 2013 to September 2014 included scaling of promising manufacturing workforce innovations and partnerships, and identifying new, concrete strategies for securing the nation's competitive advantage in transformative emerging technologies.

As a result, this report reflects not only the significant actions taken by the Advanced Manufacturing Partnership and its members to launch public-private initiatives that build on its initial recommendations, but also offers a series of further recommendations on enabling innovation in critical emerging manufacturing technologies through additional investments in innovation, securing the talent pipeline and improving the business climate for innovative manufacturing firms.

Recognizing that the U.S. manufacturing sector draws its strength from a multitude of tightly linked capabilities contributed by the private sector, academia, and labor, in its work and membership the Advanced Manufacturing Partnership has reflected the broad partnership required from across communities, educators, businesses, organized labor and government at all levels to accelerate U.S. advantage in advanced manufacturing. And through its regional working sessions and forums, the Advanced Manufacturing Partnership 2.0 Steering Committee has engaged the broader manufacturing community to highlight examples of innovative strategies that build U.S. manufacturing competitiveness.

As manufacturing grows and strengthens in the United States, a broad public-private coalition has emerged in support of American manufacturing. We have capitalized on this growing coalition and the consensus among policymakers, industry experts and academia that U.S. manufacturing matters to apply coordinated action to begin implementing the above recommendations. Already we are seeing impact from these actions - from the creation of a broad national manufacturing strategy to the launch of regional apprenticeship pilot programs. From these strong beginnings, we call for a sustained and coordinated effort to maintain momentum and engagement across the United States.

ENABLING INNOVATION

The United States' leadership in manufacturing comes from its leadership in advanced technologies and the innovation that fuels their discovery and adoption. When the United States competes in manufacturing and wins, the United States competes on the basis of this sophistication and the ability of its manufacturing industry to produce products with incredible new capabilities and functions. Sustaining U.S. competitiveness in manufacturing is thus, ultimately, an exercise in staying at the forefront of new technologies and continually breaking boundaries in both what and how it can be manufactured.

Advanced manufacturing is broadly defined and represents a continuum of interests in the manufacturing sector. Indeed, from basic metals to aerospace and electronics, manufacturing increasingly depends on advanced manufacturing components, technologies, processes, skills and strategies. To compete in the future requires a continuous transformation of manufacturing to meet the challenges posed by globalization and change.

The following recommendations build on the Advanced Manufacturing Partnership's previous report and reflect further guidance on how the United States can sustain its lead in innovation while securing critical leadership in transformative, emerging technologies:

Establishing a National Strategy for Securing U.S. Advantage in Manufacturing Technologies: Aligning on the most important technologies for U.S. competitiveness in manufacturing, and a shared vision for how to advance them, creates a platform for public-private collaboration nationally on the shared research agendas and investments required to secure and sustain a U.S. lead in these technologies. A national technology strategy outlining specific efforts and investments across the federal government and the private sector, and created and regularly updated with input from leading technologists across industry and federal labs, can optimize the nation's investment in manufacturing technology development. Critically, this national manufacturing technology strategy will provide for coordinated investments across the lifecycle of technology development, manage the investment portfolio and drive sustainable models of collaboration between the federal government and the private sector.

Over the past year, the Advanced Manufacturing Partnership 2.0 has piloted a process for developing a national manufacturing technology strategy focused on three emerging technologies of national importance – advanced sensing, controls, and platforms for manufacturing (ASCPM); visualization, informatics & digital manufacturing (VIDM); and advanced materials manufacturing (AMM) – that can serve as a model for a broader effort to develop a full national manufacturing technology strategy.

❖ **Recommendation #1**: Establish a national strategy for securing U.S. advantage in emerging manufacturing technologies with a specific national vision and set of coordinated initiatives across the public and private sectors and all stages of technology development. This should include prioritized manufacturing technology areas of national interest, leveraging the technology prioritization and analysis process developed by the Advanced Manufacturing Partnership, and should facilitate management of the portfolio of advanced manufacturing technology investments.

Coordinating Public and Private Investment in the Development of Top Emerging Manufacturing Technologies: From the experience of piloting a process for developing a national strategy to secure domestic advantage in just a subset of the important manufacturing technologies being developed today, the Advanced Manufacturing Partnership recognizes the value of regular and sustained communication and research coordination across the public and private sectors, as well as how valuable the access to top industry technologists can be for the federal government as it develops its own efforts to advance U.S. strengths in pre-competitive manufacturing technology. Building off of the increased interagency coordination on manufacturing generally made possible by the Advanced Manufacturing National Program Office, which the Advanced Manufacturing Partnership recommended in our prior report, we recommend that a continuous mechanism for research coordination across the public and private sectors, with expert input from industry and academia, be established. An Advanced Manufacturing Advisory Consortium can provide detailed, coordinated input on nascent opportunities and priorities in manufacturing that can shape national U.S. technology priorities and investments at all stages of technology development. This group, modeled after or using existing mechanisms such as the external advisory groups to the Executive Office of the President's National Science & Technology Council, could interface regularly with a standing interagency manufacturing R&D coordinating body comprising the key research and development agencies to provide feedback and partnership in the federal government's research and development priorities in advanced manufacturing. In this way, strategies for advanced manufacturing technologies could be linked to an R&D strategy.

> ❖ **Recommendation #2**: Create an Advanced Manufacturing Advisory Consortium to provide coordinated private-sector input on national advanced manufacturing technology research and development priorities.

In addition, in its pilot efforts to prioritize and develop recommendations for specific manufacturing technology areas, the Advanced Manufacturing Partnership identified a cross-cutting need for two new public-private research and technology efforts to spur the further development and adoption of these emerging technologies: the need for additional research and development infrastructure in the form of manufacturing centers of excellence (MCEs) to create a pipeline of earlier-stage technologies that can feed into the National Network for Manufacturing Innovation Institutes; and the importance of manufacturing technology testbeds (MTTs) that can de-risk the adoption of these emerging technologies, particularly for smaller manufacturers. The Advanced Manufacturing Partnership also noted the importance of security at the interface between cyber systems and physical manufacturing equipment.

In addition to fully building out the National Network for Manufacturing Innovation (NNMI), the Advanced Manufacturing Partnership recommends the creation of manufacturing centers of excellence (MCEs) that can advance earlier-stage technologies. Manufacturing centers of excellence are research laboratories, funded and operated jointly by industry and universities, to invest in basic research that responds to a particular manufacturing challenge, such as critical materials reprocessing or bonding of composite structures. These manufacturing centers of excellence can leverage existing, successful U.S. research center models co-funded by federal agencies and industry, such as National Science Foundation Engineering Research Centers. New, and potentially existing, centers within this program and other agencies' center programs could be focused on advanced manufacturing R&D as manufacturing centers of excellence, supporting existing or potential Manufacturing Innovation Institutes. These manufacturing centers of excellence could be co-located within U.S. regions shared with related Manufacturing Innovation Institutes, when feasible and advantageous to accelerated manufacturing technology maturation within the institutes. Manufacturing technology testbeds, which provide access to equipment and facilities designed for the testing and demonstration of new technologies, will enable evaluation, development, demonstration, and customization services to small, medium, and large enterprises, and vendors for technologies that are at later stages of development. They help de-risk the implementation of available technologies and help develop a talent and knowledge base for the technology or sector. Manufacturing technology testbeds are particularly needed to drive adoption of advanced technologies such as sensing, process control and IT platforms that can increase the efficiency of energy and materials use in advanced manufacturing. They can also be useful for projects that demonstrate a new production concept, increased energy or resource efficiency, or to validate a production technology against an industry standard. Manufacturing technology testbeds can also be leveraged as shared facilities with appropriate technical (computing and tool) and talent infrastructure, enabling small- and medium-sized enterprises (SMEs) to adopt new technologies and increase rapid value creation which is not possible without the availability of these shared facilities.

❖ **Recommendation #3:** Establish a new public-private manufacturing research and development infrastructure to support the innovation pipeline, which complements Manufacturing Innovation Institutes at earlier and later technology maturation stages, through the creation of manufacturing centers of excellence (MCEs) and manufacturing technology testbeds (MTTs) to provide a framework that supports manufacturing innovation at different stages of maturity and allows small and medium-sized enterprises to benefit from these investments.

Development of standards reduces the risks for enterprises developing solutions and for those implementing them, accelerating adoption of new manufactured products and manufacturing methods. The federal government should work with private industry to establish standards and interoperability for manufacturing new products and processes. This effort includes standards related to digital data, with an aim of data interoperability among systems that can speed technology adoption. Examples of such data transferability standards include communications protocols, metadata description languages, and data formats. Additionally, this effort addresses addresses the advantage of standardization for components that are physically substitutable – including information exchange standards related to materials and manufacturing processes – in focused sector areas that facilitate industry adoption of innovations by established or new manufacturers. This effort further includes cybersecurity process certification in manufacturing, not unlike ISO certification of other manufacturing processes, that can mitigate security risks at the interface of the cyber systems and physical equipment in the manufacturing ecosystem.

- ❖ **Recommendation #4:** Develop processes and standards enabling interoperability of manufacturing technologies; exchange of materials and manufacturing process information; and certification of cybersecurity processes for developers of systems.

Establishing and Supporting the National Network for Manufacturing Innovation (NNMI): In its initial report, the Advanced Manufacturing Partnership called for the creation of a National Network for Manufacturing Innovation (NNMI) to spur public-private collaborative research to address large, cross-cutting technology challenges for later stage technologies. The Advanced Manufacturing Partnership supports the Administration's actions thus far to launch four manufacturing institutes (Table A1 of Appendix A) addressing critical manufacturing technologies such as advanced composites, digital manufacturing, and lightweight metals, with four more manufacturing institutes on the way. In support of these nascent efforts to develop the National Network for Manufacturing Innovation and in anticipation of bipartisan legislation to formally establish the program, the Advanced Manufacturing Partnership recommends the development of a shared National Network for Manufacturing Innovation Governance Structure to help ensure a return on investment for Manufacturing Innovation Institute stakeholders. This governance structure should be established through clear, written guidelines covering both network governance and institute topic selection, and these guidelines should reflect multiple future scenarios for the National Network for Manufacturing Innovation: a scenario in which it is fully authorized and a scenario in which it continues to be developed solely through executive action that uses existing funds of federal agencies such as the Department of Defense and Department of Energy. The Advanced Manufacturing Partnership proposes a governance structure that maintains autonomy for individual institute operations while creating a public-private network governing council that oversees the broader performance of the network and helps to ensure maximum efficiency, collaboration and sustainability of the individual institutes.

- ❖ **Recommendation #5**: Create – through the National Economic Council, the Office of Science and Technology Policy, and the implementing agencies and departments – a shared National Network for Manufacturing Innovation (NNMI) governance structure that can ensure a return on investment for the NNMI's many stakeholders by including input from various agencies as well as private sector experts, organized labor and academia.

SECURING THE TALENT PIPELINE

Simply said, global businesses invest where the talent exists. The 2014 Manufacturing Institute survey, "Out of Inventory," reports that 75 percent of manufacturers surveyed are impacted negatively by skills shortages. Technological developments in the manufacturing sector have outpaced workforce skills, and demographic shifts have combined to create a gap in the workforce the manufacturing sector needs. The Manufacturing Institute notes that the hardest jobs to fill are those that have the biggest impact on performance, that manufacturers depend on outdated approaches for finding the right people and developing their employees' skills, that the changing nature of manufacturing work is making it harder for talent to keep up, and that the widening skills gap is expected to take the biggest toll on skilled production jobs.

The Advanced Manufacturing Partnership recognizes that in order to accelerate manufacturing innovation and growth in the United States, we must focus on developing a talented and committed workforce, and that the way in which that will happen is through public-private partnerships. Manufacturers, organized labor and academia also realize that identifying and responding to technological innovation can be expensive and time-consuming for any single firm, and thus it will be necessary that these opportunities are best addressed through partnerships of committed and motivated groups of firms and educational institutions. The challenge manufacturing firms face in filling the talent pipeline is compounded by the misperceptions the general public has about careers in manufacturing, and AMP2.0 addressed this critical aspect to help secure the sector's future workforce.

Shifting the Misconceptions the Public Holds of Manufacturing: For decades, workers flocked to manufacturing careers because those jobs were viewed as stable, solid careers that provided a path to the middle class for workers at every educational level. That belief has been shaken by the job losses of the past decades. The first step in securing the talent pipeline for manufacturing is to shift how manufacturing careers are perceived, and this will require a focused and sustained national and local effort. AMP 2.0 is partnering with private industry and media to create and launch a series of advertisements and events, including National Manufacturing Day, to convey the current excitement about and opportunities in manufacturing careers.

The AMP2.0 SC recommends that the federal government launch a national campaign to change the image of manufacturing, and also undertake activities to support National Manufacturing Day. The AMP2.0 recommends that efforts to change the image of manufacturing include efforts modeled on exemplary programs that strive to interest students in advanced manufacturing and product realization starting at early ages and through graduate education.

❖ **Recommendation #6:** <u>Launch a national campaign to change the image of manufacturing, and support National Manufacturing Day's efforts to showcase real careers in today's manufacturing.</u>

Connecting More Americans with Skills for Successful Careers in Manufacturing: Building on its recommendations in its first report, the Advanced Manufacturing Partnership has also taken new action to secure the talent pipeline and connect more Americans with the skills needed for successful careers in manufacturing. These efforts have been focused in four areas: developing a national system of skills certifications and accreditation, tapping the talent pool of returning veterans, investing in community colleges by building successful apprenticeships to deliver demand-driven training, and documenting best practices in the development of career pathways in advanced manufacturing education.

Credentials and accreditations are the outcomes of many years of skilled trades training and hands-on learning through apprenticeships. These professional credentials are a gateway to higher wages, increased career mobility and provide an employee the ability to follow the job market. The relevance of credentials and accreditations can vary regionally, or by employer, and is of course highly dependent on the specific job opening. The Advanced Manufacturing Partnership, as it recommended in its first report, encourages the partial funding and implementation of a system of nationally recognized, portable, and stackable skill certifications that employers can preferentially utilize in hiring and promotion.

To help advance the development of a national system of skills certification and accreditation, the Advanced Manufacturing Partnership has developed a template for a "pathways" model for advanced manufacturing training and education with multiple on- and off-ramps and multiple stackable completion options including industry-recognized certificates, formal diplomas, and college degrees. It is also important to include on-line training as a key component of these on- and off-ramps, recognizing that new and mature students require flexibility and convenience as they pursue skill and career development. These certifications and online training platforms can leverage prior investments in the Department of Labor Trade Adjustment Assistance-Community College and Career Training (TAACCCT).

❖ **Recommendation #7:** <u>Incent private investment in the implementation of a system of nationally recognized, portable, and stackable skill certifications that employers utilize in hiring and promotion, by providing additional funds that build on investments being made through the Department of Labor and Department of Education Trade Adjustment Assistance Community College and Career Training (TAACCCT).</u>

As the Advanced Manufacturing Partnership recognized in its first report, the Community College level of education is the 'sweet spot' for impact on the skills gap in manufacturing. Increased investment in this sector was recommended by the Advanced Manufacturing Partnership two years ago, following the best practices of leading innovators. Because apprenticeships are so integral to a tradesman earning a professional credential or accreditation, and because hands-on learning is of utmost importance for many careers in advanced manufacturing, the Advanced Manufacturing Partnership organized and piloted an apprenticeship program with three major employers and two colleges, and documented the approach in a playbook such that the model can be further replicated. This pilot program illustrated the importance of collaboration between community colleges and employers to deliver customized training that meets the unique needs of employers and provides flexibility to future employees. AMP also formed a statewide consortium of 12 colleges, 2 Manufacturing Centers of Excellence and more than 25 manufacturers in Minnesota to launch the Learn, Work, Earn model to align trainings and fill jobs.

Finally, the Advanced Manufacturing Partnership identified examples of exemplary Career Education Pathways from other industries in order to build a pathway model for Advanced Manufacturing, also recognizing the need for flexible workforce training solutions so that both new and mature employees can develop their skills in an individualized way. AMP2.0 provided recommendations on how to duplicate, scale-up and improve on the best programs. Across all of these exemplary career pathways, common elements emerged, including the use of nationally recognized, portable, and stackable skills certifications as a tool for signaling skills mastery to employers, and the increasing use of online training platforms to increase access to and the scale of training programs.

- **Recommendation #8:** Make the development of online training and accreditation programs eligible to receive federal support, for example through federal jobs training programs.

In addition, returning veterans possess many of the key skills needed to fill the skills gap in the manufacturing talent pipeline. With more than 400,000 veterans leaving the armed forces for other opportunities in the next two years, AMP2.0 recognizes that tapping this dedicated, driven talent pool can help alleviate the skills shortage manufacturers face *today*, and in most cases without the need for significant additional training. There are two barriers to a smooth transition from the military to manufacturing: little understanding of today's manufacturing careers, and difficulty equating military skills with private sector job qualifications. In order to help connect veterans with employers, AMP2.0 completed an inventory of veteran resources, wrote practical guides for veterans, employers and academic institutions to help transition veterans to careers in manufacturing, and provided recommendations for skills translators and Veterans' Skills Badging programs.

To ensure the sustainability and accessibility of the workforce development best practice tools created by the Advanced Manufacturing Partnership, the Manufacturing Institute has volunteered its resources and platform to curate the documents, toolkits and playbooks that have been created by the Advanced Manufacturing Partnership, helping to further scale and replicate these important talent development opportunities.

> ❖ **Recommendation #9**: <u>Curate the documents, toolkits and playbooks that have been created by AMP2.0 to further scale and replicate these important talent development opportunities, via the Manufacturing Institute.</u>

Finally, although AMP's actions and recommendations on workforce training focused chiefly on strengthening the role of community colleges within the pipeline, AMP2.0 also recognizes unique needs and opportunities for four-year colleges, universities, and graduate programs in educating leaders who develop and implement advanced manufacturing technology. Most U.S. undergraduate engineering programs are reviewed by the Accreditation Board of Engineering and Technology (ABET). ABET should consider potential changes in the engineering curriculum requirements that include advanced manufacturing skills. Further, U.S. graduate programs (M.S., Ph.D., and M.B.A.) should consider specific skillsets and areas of high interest and demand in manufacturing, including opportunities for the public and private funding of graduate fellowships as noted in Appendix 1.

IMPROVING THE BUSINESS CLIMATE

As countries compete for the advanced manufacturing industries that provide the foundation for future innovation, it is critical that the United States offer an environment where young companies are able to demonstrate the viability of new technologies at scale and where mature, main street manufacturers are able to access the capital and capabilities they need to expand into new business opportunities. Unlike services and software, manufacturing requires unique capital that often cannot be rapidly brought on line or redeployed for other uses. Long time lines, higher technology risk, and large capital requirements combine to create a risk profile unacceptable to many investors. In addition, many connections between manufacturers and the capabilities, know-how, and capital for expansion need to be rebuilt after the last decade of lost growth in the manufacturing sector.

Expanding and Enhancing Intermediary Solutions for Manufacturing Expertise: Small and medium manufacturers can adopt and leverage manufacturing technologies more readily when improved access to information is provided via "intermediary services" – established organizations that understand and facilitate access to a wide range of information needed for technology commercialization and scale up, including technical expertise, supply chain partners, financing options, and government programs. AMP2.0 provided three key design characteristics for successful intermediaries including 1) regional scope, 2) technology or industry specific focus, and 3) service provision to a network of firms rather than a single firm. The Manufacturing Extension Program is one example of an intermediary that could be enhanced or configured to meet these key design criteria. In addition, accurate market and technical insight is critical for small and medium manufacturers to develop an entry strategy, and mobilize resources to adopt new technologies for processes, materials, and new products. Shared facilities with effective technical and talent, including the manufacturing technology testbeds, will also enable digital design and manufacturing to be realized by networks of small and medium-sized enterprises. AMP2.0 conducted a pilot project to outline an approach for small and medium-sized manufacturers to access market and technical insights required to scale-up new technology. Tools developed from the pilot could be utilized by an intermediary to support small and medium-sized manufacturers in developing a business plan for scale-up.

- ❖ **Recommendation #10**: <u>Leverage and coordinate existing federal, state, industry group and private intermediary organizations to improve information flow about technologies, markets and supply chains to small and medium-sized manufacturers.</u>

Increasing capital access for established and start-up firms: Advanced manufacturing small and medium-sized manufacturers (SMMs) often are not compelling investments for capital markets due to technology risk, market adoption risk, long lead times to significant revenue and significant capital requirements. Reducing capital requirements is one way that government has encouraged manufacturers through grants, loan guarantees and tax deferrals. Frequently overlooked are the other means that investments can be made more attractive to capital markets through demand creation, reducing technical risk, and reducing development time. For example, by offering low-cost loans to private-sector investors in "first-of-a kind" production facilities, a public-private Scale-Up Fund could incentivize additional investment in novel production facilities. Similar incentives at the regional and national levels can also help create a vibrant, domestic equipment supply base for specific nascent technologies such as in additive manufacturing. In addition, existing federal programs and authorities can use the preannouncement of their future demand to reduce the market risk of new technologies, and thereby increase their attractiveness for investment.

❖ **Recommendation #11**: Reduce the risk associated with scale-up of advanced manufacturing by improving access to capital through the creation of a public-private scale-up investment fund; the improvement in information flow between strategic partners, government and manufacturers; and the use of tax incentives to foster manufacturing investments.

In addition to its efforts this year to further develop and implement its recommendations on critical emerging technology areas, securing the talent-pipeline, and creating a supportive environment for innovative manufacturing businesses, the Advanced Manufacturing Partnership reiterates the recommendations on tax and regulatory policy advanced in its original report. The Advanced Manufacturing Partnership supports the Administration's efforts regarding retrospective review to streamline regulations to create a more competitive environment for manufacturing. The Advanced Manufacturing Partnership also emphasizes the importance of passing fundamental business tax reform while deepening those incentives designed to encourage the long-term establishment of capital-intensive and space-intensive manufacturing operations as part of coordinated national strategies to become manufacturing leaders in a specific sector.

IMPLEMENTATION

Ongoing efforts over the last five years have required substantial coordination across federal agencies and collaboration with the private sector. To ensure successful implementation of the above recommendations addressed to the federal government, sustained leadership from the Executive Office of the President (EOP) is a necessity. The Office of Science and Technology Policy (OSTP) and the National Economic Council (NEC) have naturally played central roles in moving multi-agency initiatives forward, including the implementation of key recommendation from previous PCAST reports on advanced manufacturing. In addition to EOP leadership and deep engagement, clear accountability across multi-agency initiatives is critical to accelerate needed outcomes.

❖ **Recommendation #12**: The National Economic Council (NEC) and the Office of the Science and Technology Policy (OSTP), within 60 days, should submit to the President a set of recommendations that specify: (1) the ongoing EOP role in coordinating the federal government's advanced manufacturing activities; and (2) clear roles and responsibilities for federal agencies and other federal bodies in implementing the above recommendations.

I. Advanced Manufacturing Partnership 2.0

Introduction

The United States has long thrived as a result of its ability to manufacture goods and sell them to global markets. Manufacturing supports the country's economic growth, leading the Nation's exports and employing millions of Americans. In addition, manufacturing drives knowledge production and innovation in the United States by supporting two-thirds of private sector research and development and by employing the vast majority of U.S. scientists, engineers, and technicians to invent and produce new products[1].

The United States has been the leading producer of manufactured goods for more than 100 years. Yet, in the 2000's, manufacturing faced major employment declines as factories were shuttered. U.S. strengths in manufacturing innovation and technologies that have sustained American leadership in manufacturing are under threat from new and growing competition abroad.

U.S. strengths in manufacturing innovation and invention that have sustained American leadership in manufacturing remain under threat from new and growing competition abroad.

The United States has been losing significant elements of research and development linked to manufacturing, as well as the ability to compete in the manufacture of many products that were invented and innovated here. And other countries have been stepping up their investments in manufacturing and in research, eroding the United States' lead.

Beginning in 2011, the United States began a series of national-level discussions and actions between the public and private sectors to ensure this country is prepared to lead the next generation of manufacturing. Those activities, which came to be called the Advanced Manufacturing Partnership, promise a new era of manufacturing in this country.

This report describes the Advanced Manufacturing Partnership 2.0 (AMP2.0)'s actions and recommendations for accelerating progress along three integrated pillars that together can strengthen the U.S. ecosystem for advanced manufacturing leadership: Enabling Innovation, Securing the Talent Pipeline, and Improving the Business Climate. These recommendations focus on both federal actions and public-private partnerships that can accelerate U.S.-based

---

[1] President's Council of Advisors on Science and Technology, *Ensuring America's Leadership in Advanced Manufacturing*, June 2011

manufacturing across a wide range of industry sectors. The United States is poised to lead this manufacturing innovation by leveraging regional strengths throughout the country. However, to do so, the United States must implement a sustained and coordinated national effort to grow our lead in innovation, to develop the skills needed in today's advanced manufacturing plants, and to increase the competitiveness of our environment for manufacturing that recognizes the stiff competition from other nations with more centralized plans and practices.

Process

In June 2011, the President's Council of Advisors on Science and Technology (PCAST) and the President's Innovation and Technology Advisory Committee (PITAC) issued the report to the President: *Ensuring American Leadership in Advanced Manufacturing*. The report outlined a strategy and specific recommendations for revitalizing the Nation's leadership in advanced manufacturing.

Based on that report, in 2011 President Obama launched the Advanced Manufacturing Partnership (AMP), a national effort bringing together industry, universities, the Federal government, and other stakeholders to identify emerging technologies with the potential to create high quality domestic manufacturing jobs and enhance U.S. global competitiveness. Operating within the framework of PCAST, the Advanced Manufacturing Partnership Steering Committee (SC) identified three pillars to ensure an ecosystem for advanced manufacturing leadership:

(1) Enabling Innovation,

(2) Securing the Talent Pipeline, and

(3) Improving the Business Climate.

As described in the *Report to the President on Capturing Domestic Competitive Advantage in Advanced Manufacturing* (July 2012), the partnership's 2012 recommendations fit intimately together and have an additive effect across these three pillars. In its first set of recommendation, recommendation, the Advanced Manufacturing Partnership sought to detail a model for evaluating, prioritizing, and recommending federal investments in advanced manufacturing technologies. It recommended public-private partnerships, including the National Network for Manufacturing Innovation, focused on advancing high-impact technologies and models for collaboration that encompass technology development, innovation infrastructure, and workforce development. It also recommended to the administration actions to increase private-private-sector investment in advancing manufacturing in the United States. Appendix A lists the

the implementation status of the recommendations made by PCAST to the Federal government through the Advanced Manufacturing Partnership report of July 2012.

In September 2013, the Advanced Manufacturing Partnership 2.0 (AMP2.0) convened the second phase of the partnership. AMP2.0 was charged by PCAST to develop specific, targeted, and actionable recommendations, building on the recommendations in the 2012 report, which would improve and sustain U.S. manufacturing innovation. Through a vibrant collaboration among industry, academia, labor organizations, and government agencies, AMP2.0 made significant progress toward this goal, recommending paths for sustained government engagement on advanced manufacturing with private organizations at local, state, regional, and national levels. AMP2.0 members further led private sector activities set out in the 2012 Advanced Manufacturing Partnership recommendations for the industrial, nonprofit, and academic manufacturing communities.

In its efforts, the AMP2.0 SC called on the expertise of 43 college and university faculty and administrators, 51 industry leaders and employees, four labor group representatives, and six independent experts for the working teams, in addition to the contributions from countless participants at regional meetings, roundtables and via other forum.

AMP 2.0 Steering Committee Actions & Recommendations

The AMP Steering Committee put forth a set of recommendations in 2012 around three pillars: Enabling Innovation, Securing the Talent Pipeline, and Improving the Business Climate. The AMP2.0 Steering Committee's actions and further recommendations are offered to accelerate manufacturing innovation and growth, ensuring global competitiveness of the U.S. manufacturing sector, and fueling the innovation economy.

The significant body of work accomplished by AMP2.0 includes the development of national strategies piloted on three of the technologies identified by the Advanced Manufacturing Partnership in its first report as technology areas of high national priority. Through piloting a process for creating national technology strategies, the Advanced Manufacturing Partnership identified a need for sustained public-private coordination channels for manufacturing technology investments and identified new forms of research and development infrastructure that can help advance a cross-cutting set of technologies. Because new advanced manufacturing technologies will require a skilled workforce, AMP2.0 has also implemented several concrete actions to accelerate jobs-driven workforce training primarily through partnerships between local academic institutions and local employers. A focus on these best practices and participation of all stakeholders (government, industry, and academia) has led to new innovations in the delivery of workforce training. Finally, the Advanced Manufacturing Partnership 2.0 has focused on those elements of improving the business climate that matter most for small and medium-sized manufacturers and to new emerging technology companies as they seek new business opportunities created by further deploying new innovations and technologies.

In addition to identifying steps the federal government can take to secure an enduring U.S. lead in advance manufacturing, the Advanced Manufacturing Partnership 2.0 also took action where it could to launch public-private initiatives building towards the same goal, and in many places it has identified additional opportunities for public-private action or private sector leadership to strengthen U.S. manufacturing. These are described more fully in Appendices 1-5 and related Annexes. For those areas where the federal government is critical to advancing U.S. competitiveness in manufacturing, a summary of the recommendations follows.

Summary of the Advanced Manufacturing Partnership 2.0's Recommendations

	Pillar I: Enabling Innovation
Recommendation #1	***Establish a national strategy for securing U.S. advantage in emerging manufacturing technologies*** with a specific national vision and set of coordinated initiatives across the public and private sectors and all stages of technology development. This should include prioritized manufacturing technology areas of national interest, leveraging the technology prioritization and analysis process developed by the Advanced Manufacturing Partnership, and should facilitate management of the portfolio of advanced manufacturing technology investments.
Recommendation #2	***Create an Advanced Manufacturing Advisory Consortium*** to provide coordinated private-sector input on national advanced manufacturing technology research and development priorities.
Recommendation #3	***Establish a new public-private manufacturing research and development infrastructure to support the innovation pipeline***, which complements Manufacturing Innovation Institutes at earlier and later technology maturation stages, through the creation of manufacturing centers of excellence (MCEs) and manufacturing technology testbeds (MTTs) to provide a framework that supports manufacturing innovation at different stages of maturity and allows small and medium-sized enterprises to benefit from these investments.
Recommendation #4	***Develop processes and standards*** enabling interoperability of manufacturing technologies; exchange of materials and manufacturing process information; and certification of cybersecurity processes for developers of systems.
Recommendation #5	***Create*** – through the National Economic Council, the Office of Science and Technology Policy, and the implementing agencies and departments – ***a shared National Network for Manufacturing Innovation (NNMI) governance structure*** that can ensure a return on investment for the NNMI's many stakeholders by including input from various agencies as well as private sector experts, organized labor and academia.

	Pillar II: Securing the Talent Pipeline
Recommendation #6	***Launch a national campaign*** to change the image of manufacturing and support National Manufacturing Day's efforts to showcase real careers in today's manufacturing sector.
Recommendation #7	***Incent private investment in the implementation of a system of nationally recognized, portable, and stackable skill certifications that employers utilize in hiring and promotion***, by providing additional funds that build on investments being made through the Department of Labor and Department of Education Trade Adjustment Assistance Community College and Career Training (TAACCCT).
Recommendation #8	***Make the development of online training and accreditation programs eligible to receive federal support*** through federal jobs training programs.
Recommendation #9	***Curate*** the documents, toolkits and playbooks that have been created by AMP2.0 to further scale and replicate these important talent development opportunities, via the Manufacturing Institute.

	Pillar III: Improving the Business Climate
Recommendation #10	***Leverage and coordinate existing federal, state, industry group and private intermediary organizations*** to improve information flow about technologies, markets and supply chains to small and medium-sized manufacturers.
Recommendation #11	***Reduce the risk associated with scale-up of advanced manufacturing*** by improving access to capital through the creation of a public-private scale-up investment fund; the improvement in information flow between strategic partners, government and manufacturers; and the use of tax incentives to foster manufacturing investments.

Each of these three pillars – enabling innovation, securing the talent pipeline, and improving the business climate – are mutually reinforcing and individually important for securing sustained U.S. leadership in advanced manufacturing and innovation. Below, the Advanced Manufacturing

Partnership 2.0 discusses in detail its findings, activities, and latest set of recommendations within each pillar for strengthening U.S. advanced manufacturing.

Implementation of these recommendations directed toward the federal government requires coordination among and action plans by federal agencies and bodies. This is addressed in the final recommendation of this report:

	Implementation
Recommendation #12	*The National Economic Council (NEC) and the Office of the Science and Technology Policy (OSTP), within 60 days, should submit to the President a set of recommendations* that specify: (1) the ongoing EOP role in coordinating the federal government's advanced manufacturing activities; and (2) clear roles and responsibilities for federal agencies and other federal bodies in implementing the above recommendations.

II. Recommendations

Pillar 1: Enabling Innovation

Leadership in innovation and manufacturing technologies can cement the basis for sustained U.S. competitiveness in manufacturing. However, maintaining that leadership position requires a clear focus and coordinated effort to invest in technologies of national priority. Technology breakthroughs that can take innovations out of a lab and into production, require shared infrastructure to support cross-cutting discoveries. Not least among this shared infrastructure is the National Network for Manufacturing Innovation.

Focusing resources on priority opportunities to develop technologies that can sustain U.S. leadership in advanced manufacturing is the first important step in accelerating innovation in manufacturing technology. AMP2.0 developed and implemented a process by which several potential high-impact technologies were prioritized. Three technologies were elevated for further analysis by industry, organized labor and academic experts based on their potential for cross-cutting impact on U.S. advanced manufacturing industries, broad private sector pull, importance to national security, and their ability to build long-term U.S. competitiveness in advanced manufacturing: Advanced Sensing, Control, and Platforms for Manufacturing (ASCPM); Visualization, Informatics and Digital Manufacturing Technologies (VIDM); and Advanced Materials Manufacturing (AMM).

Because the discoveries and investments needed to build a U.S. advantage in these technologies are often beyond the resources of one firm and benefit from insights from multiple disciplines and industries, shared infrastructure is a key enabler to support and promote the efficient development of new technologies. At earlier stages of development, co-investments in basic research for manufacturing applications can build a pipeline of discoveries leading to real technological advancement and, for those technologies that are ready to deploy, technology test beds can help de-risk the adoption of these technologies, especially for small and medium-sized manufacturers, helping ensure that cutting-edge technologies make their way onto U.S. factory floors. Industry-developed standards play a similar role in helping speed the adoption of new technologies.

The National Network for Manufacturing Innovation, recommended in the Advanced Manufacturing Partnership's first report, provides critical shared infrastructure to take technologies from research to late-stage technologies that are close to deployment. The creation of four pilot manufacturing institutes and the pledge to launch four more this year represent a significant step towards creating the National Network for Manufacturing Innovation and addressing the gap in shared manufacturing research infrastructure. Already, the pilot manufacturing institutes are illustrating the advantages of public-private partnerships and shared infrastructure to streamline and de-risk the process by which new manufacturing technologies are made real. It is critical, however, to ensure the long-term return on investment of these manufacturing institutes, and this is best done through the creation and implementation of standard principles for the protection of intellectual property and a strong governance body.

Establishing a National Manufacturing Technology Strategy

The research and innovation ecosystem of a nation is highly dependent on the presence of a manufacturing base that provides constant feedback in terms of technology problems and challenges to be solved.

Since technology is rapidly and continually advancing, one of the key goals of the Advanced Manufacturing Partnership was to develop and establish a permanent mechanism to identify the next generation of advanced manufacturing technologies that will have the greatest impact on the growth and competitiveness of the United States, and to provide technical insight into the opportunities and obstacles each critical technology will face as it progresses towards commercialization.

Unlike the United States today, many leading industrialized countries follow a systematic prioritization and planning process that is explicitly aligned to their national interests and strategies. For the United States to benefit from a similar exercise in defining a national vision for advanced manufacturing technologies and a shared set of research and development priorities to advance them, any national prioritization effort in the United States must build on the strengths of our decentralized research and development efforts and industrial base. An approach to developing national technology strategies in the United States should capitalize on the flexibility of the U.S. academic-industrial partnership afforded by technical and community colleges, universities, and national and military research laboratories. These public-private partnerships, with input from the workforce itself, can accelerate the efficient and cost-effective commercialization of new technologies by de-risking the investment during development.

The three interrelated mechanisms proposed below – an advisory mechanism, coordination of advanced manufacturing R&D, and a strategy process – will be critical parts of the national strategy and could help ensure a highly effective public-private partnership for advanced manufacturing upon conclusion of the AMP2.0 effort.

In order to support a relevant, sustainable and transparent national strategy for advanced manufacturing, AMP2.0 used three key manufacturing technology areas as examples of the power and potential of scoping and prioritizing technology strategies. As detailed in Appendix 1, these areas were prioritized among the 11 MTAs listed in the AMP 2012 report, using four criteria:

1. Industry/market pull
2. Cross-cutting impact across multiple industry sectors
3. Importance to national security and competitiveness
4. Leverages current US strengths/competencies

This process analyzes the current technology readiness, timeline to advance the technology, and the gaps or obstacles that must be addressed. This analysis can, and should, inform the national manufacturing strategy and portfolio prioritization within and among federal agencies, and be kept relevant with periodic reassessments.

Using this prioritization and technology assessment process, AMP2.0 created strategies for three transformative manufacturing technologies: Advanced Sensing, Control, and Platforms for Manufacturing (ASCPM); Visualization, Informatics and Digital Manufacturing Technologies (VIDM); and Advanced Materials Manufacturing (AMM).

Appendix 1 includes a high-level summary of this effort, and Annexes 1-10 provide nine letter reports detailing the technology gaps and recommendations for the three Manufacturing Technology Areas (MTAs) that AMP2.0 prioritized for this assessment. The findings from the analyses for each of these MTAs are noted in Table 1. These three MTAs are described briefly as:

- **Advanced Sensing, Control, and Platforms for Manufacturing (ASCPM)**: A new generation of network-based information technologies has created access to new uses of data and information as new products and manufacturing methods are developed. These technologies make a seamless interaction between cyber and physical assets possible. The research in this space is focused on embedded sensing, measurement and control systems with scalable IT platforms.

- **Visualization, Informatics and Digital Manufacturing Technologies (VIDM):** This technology is important as researchers and manufacturers move from digital design, to planning, to purchasing and delivery of raw materials, and finally to the manufacture of customized products. One aspect of the technology deals with supply chain efficiency, and the other aspect deals with the speed with which products are designed, manufactured and brought to market. The research in this space is focused on embedded sensing, measurement and control systems into materials and technologies. When this link is strong, it increases productivity, product and process agility, environmental sustainability, improved energy and raw material usage, better safety performance and improved economics.

- **Advanced Materials Manufacturing (AMM):** Novel new materials are being designed at a quickened pace over the last decade due to better modeling technology and high-throughput research. Materials innovation is a key to U.S. competitiveness, given the historic national security implications of unstable supplies of important materials, and their improved environmental profile over traditional materials. AMM is focused on the design and synthesis of new materials, as well as innovative approaches to processing of traditional materials. Because the materials are so different, AMP2.0 has prepared three additional letters, each analyzing one specific aspect of advanced materials manufacturing: (1) advanced structural composites; and (2) biomanufacturing [of biological therapeutics]; and (3) critical materials reprocessing. See Appendix 1 for drivers motivating analysis of manufacturing in these three areas.

Table 1. AMP2.0 technology strategy recommendations for three prioritized Manufacturing Technology Areas.

Technology areas:	Advanced Sensing, Control, and Platforms for Manufacturing	Visualization, Informatics and Digital Manufacturing	Advanced Materials Manufacturing
R&D Infrastructure to Support the Innovation Pipeline	▪ Establish Manufacturing Technology Testbeds (MTTs) to demonstrate the use of and business case for new technologies, including "smart manufacturing" capabilities.	▪ Create a Manufacturing Center of Excellence (MCE), focused on basic research at earlier technology development levels, on the Digital Thread, including tools for digital design and energy efficient digital manufacturing.	▪ Launch Materials Manufacturing Centers of Excellence (MCEs) to support R&D in topics that support MIIs and other manufacturing technology areas in the national strategy.

The National Network for Manufacturing Innovation	▪ Establish an institute focused on ASCPM for energy use optimization in energy-intensive and digital information-intensive manufacturing.	▪ Launch a Big Data MII focused on secure analysis of and decision-making via large, integrated data sets for manufacturing processes (in addition to the current Digital Manufacturing and Design Innovation Institute).	▪ Leverage supply chain management of defense assets to spur innovation and RD&D in critical materials reprocessing.
Public-Private Technology Standards	▪ Develop new industry standards, including data interoperability standards for key systems and vendor support.	▪ Craft and deploy policy standards for manufacturing cyber-physical security and digital data exchange and ontology.	▪ Design data standards for material characterization to enable rapid uptake of new materials and manufacturing methods
Additional Strategies		▪ Incentivize creation and commercialization of additive manufacturing systems providers, service bureaus or system integrators.	▪ Establish Manufacturing Innovation Fellowships for Ph.D.'s in key AMM areas, such as biotherapeutic manufacturing.

AMP2.0's process demonstrated that the following considerations critically impact technology strategy development and deployment. First, the technology vision must include technologies at

different stages of maturity. Second, equal consideration must be given to the technology push from academia and government research, and to the market-pull perspectives of industry and other agencies. Third, the engagement of interagency experts and research leaders would provide valuable information to inform the national strategy. Fourth, many government mechanisms such as the Manufacturing Extension Partnership and M-TAC pilots being initiated by NIST exist to drive manufacturing technology development and should be considered and leveraged when appropriate to implement strategies.

The three pilot technology strategies developed by AMP2.0 proved the usefulness of a public-private partnership to inform the prioritization and analysis of key technologies. For the pilot technology strategies, expert teams focused on individual technologies while enlisting broad engagement of experts from companies and universities alongside input from federal research and development agencies. Along the way, this engagement process spurred additional integration and alignment across the individual technology development efforts of companies, universities, and federal agencies The Advanced Manufacturing Partnership 2.0's technology strategy development process also made clear the importance of coordination across the federal research and development agencies, given their roles in advancing different stages or aspects of any individual technology.

> ❖ **Recommendation #1:** <u>Establish a national strategy for securing U.S. advantage in emerging manufacturing technologies with a specific national vision and set of coordinated initiatives across the public and private sectors and all stages of technology development. This should include prioritized manufacturing technology areas of national interest, leveraging the technology prioritization and analysis process developed by the Advanced Manufacturing Partnership, and should facilitate management of the portfolio of advanced manufacturing technology investments.</u>

The pilot national strategies in these three technology areas that emerged, and are detailed in the appendix and annexes to this report, identified opportunities for private partners and many existing federal programs to coordinate public and private investments in addressing significant research questions that need to be solved for an individual technology to advance.

Going forward, the Advanced Manufacturing Partnership recommends the creation of a standing process to develop national technology strategies in advance manufacturing and of a mechanism to continually engage the private sector for its insights on the state of technology and to enable the coordination of public and private investments in technology development. In addition to leveraging the improved interagency coordination on manufacturing initiatives spurred by the Advanced Manufacturing National Program Office, the creation of national technology strategies and R&D coordination in manufacturing could be modeled off of the interagency process currently used by the Executive Office of the President's National Science and Technology Committee which currently crafts national research and development strategies in a range of technology areas. In addition, the Advanced Manufacturing Partnership 2.0 recommends that the federal government create an Advanced Manufacturing Advisory Consortium, leveraging leading technologists from across industry and universities, to provide a continuous channel for the federal government to access private sector insights in crafting the national technology strategies and to ensure ongoing coordination of public and private investments.

> ❖ **Recommendation #2:** <u>Create an Advanced Manufacturing Advisory Consortium to provide coordinated private-sector input on national advanced manufacturing technology research and development priorities.</u>

In addition to research areas that could benefit from coordinated public and private investment, in each technology area, the expert teams identified a similar need for shared, public-private research and development infrastructure to help advance more industry-focused basic research at the front-end of the technology development pipeline, to harness efforts across industry, universities, and agencies to address key technology challenges at later stages of development, and, once technologies were available, to de-risk their adoption on U.S. factory floors, especially for small and medium sized manufacturers.

> ❖ **Recommendation #3:** <u>Establish a new public-private manufacturing research and development infrastructure to support the innovation pipeline, which complements Manufacturing Innovation Institutes at earlier and later technology maturation stages, through the creation of manufacturing centers of excellence (MCEs) and manufacturing technology testbeds (MTTs) to provide a framework that supports manufacturing innovation at different stages of maturity and allows small and medium-sized enterprises to benefit from these investments.</u>

To spur more industry-focused basic research in advanced manufacturing at the front-end of the innovation pipeline, the Advanced Manufacturing Partnership 2.0 recommends the creation of Manufacturing Centers of Excellence, basic research laboratories jointly funded and operated by industry and universities addressing targeted industry technology needs. In many cases, these basic research centers can ensure that there are sufficient early-stage manufacturing technologies that can ultimately feed in to the later-stage development and deployment driven by the National Network for Manufacturing Innovation.

For example, in visualization, informatics, and digital manufacturing, the Advanced Manufacturing Partnership 2.0 expert teams identified a need for a manufacturing center of excellence focused on improving the capabilities of technologies, primarily software, which connects the "digital thread" or secure flow of digitized information between design, simulation, and production stages. The expert team investigating advanced materials manufacturing identified a need for manufacturing centers of excellence in materials processing science and manufacturing engineering needed for industrial applications, such as composites joining or critical materials reprocessing. In both instances, these manufacturing centers of excellence would be investing in basic research "upstream" from existing or new manufacturing innovation institutes that are envisioned to comprise the National Network for Manufacturing Innovation.

Historically, the National Science Foundation has funded Engineering Research Centers in advanced manufacturing fields that have united $3-4 million in federal funds for engineering research at leading universities with significant co-investments from a broad range of industry partners. The Advanced Manufacturing Partnership believes that this model provides a useful template for the design, operations, and creation of Manufacturing Centers of Excellence. New, and potentially existing, centers within this program and within other agencies' center programs could be focused on advanced manufacturing R&D as manufacturing centers of excellence, supporting existing or potential Manufacturing Innovation Institutes. Here, we note that manufacturing centers of excellence provide R&D at early technology readiness level/manufacturing readiness levels, and can thus include manufacturing technology areas not yet sufficiently mature for or prioritized as Manufacturing Innovation Institutes (MIIs). Likewise, manufacturing centers of excellence initiated to support existing MIIs provide a stream of early-stage technology innovation as input for MIIs that are, by design, focused at higher technology/manufacturing readiness levels. Manufacturing centers of excellence should be co-located within U.S. regions shared with related Manufacturing Innovation Institutes, when feasible and advantageous to accelerated manufacturing technology maturation within the institutes. Here, the definition of "region" can vary broadly among manufacturing technology areas, and can include corridors spanning multiple geographically contiguous states, as well as virtual regions of related manufacturing capabilities, industries, and workforce.

At the opposite end of the technology development pipeline, many new technologies that are already available for deployment face slow adoption, as individual manufacturers struggle to quantify the value of those new technologies compared with the risk of being an early adopter. To help de-risk the adoption of available technologies – for example, many technologies in the advanced sensing, control, and platforms area – manufacturing technology testbeds (MTTs) can provide shared equipment and infrastructure for manufacturers to demonstrate, evaluate, and explore customizing new technologies. Especially for small- and medium-sized manufacturers (SMMs) who can rarely afford to build their own testing infrastructure – and for whom a technology failure could be more catastrophic –these manufacturing technology testbeds can significantly de-risk the implementation of available technologies and also contribute to the development of a talent and knowledge base in the use of that technology. And in many cases, manufacturing technology testbeds can lead to further insights about the use of a technology in production or help validate a production technology against an industry standard. While the range of federal investment in a manufacturing technology testbed varies across manufacturing technology areas, in general, a manufacturing technology testbed can be created with $5 to $10 million of federal or state funds matched by an equivalent amount or more of private sector investment.

AMP2.0 emphasizes that the envisioned role of the manufacturing centers of excellence and manufacturing technology testbeds is primarily to support the early-stage and late-stage readiness levels of manufacturing technologies developed within Manufacturing Innovation Institutes. Coordination and leveraging across this innovation pipeline that comprises research, development, demonstration, and deployment is expected. At the same time, the manufacturing centers of excellence also provide the United States a coordinated means to remain at the cutting edge of manufacturing technology innovation, including in emergent areas that may later be identified and prioritized as Manufacturing Innovation Institutes; while the manufacturing technology testbeds address a critical need for small and medium-sized manufacturers to test and adapt matured technologies including but not limited to Manufacturing Innovation Institute output.

Finally, across all three pilot technology strategies, the Advanced Manufacturing Partnership 2.0's expert teams identified a need for new industry-driven standards to help spur the adoption of new technologies, products and manufacturing methods. Standards allow a more dynamic and competitive marketplace, without hampering the opportunity to differentiate. Development of standards reduces the risks for enterprises developing solutions and for those implementing them, accelerating adoption of new manufactured products and manufacturing methods. The federal government should work with private industry to establish standards and interoperability for manufacturing new products and processes. This includes data standards for the interoperability of manufacturing hardware and software to speed the adoption of new digital manufacturing techniques, the use of advanced sensing, controls, and platform technologies, and new cybersecurity standards in manufacturing that can mitigate security risks at the interface of the cyber systems and physical equipment in the manufacturing ecosystem, important again for the development of digital manufacturing and the use of advanced sensors.

❖ **Recommendation #4:** <u>Develop processes and standards enabling interoperability of manufacturing technologies; exchange of materials and manufacturing process information; and certification of cybersecurity processes for developers of systems.</u>

Establishing & Supporting the National Network for Manufacturing Innovation Institutes

In its first report, the Advanced Manufacturing Partnership recognized the need for a National Network for Manufacturing Innovation (NNMI), composed of institutes that would represent a long-term partnership between industry and universities, enabled by local, state and federal government dedicated to addressing critical development and deployment challenges for later-stage technologies. The National Network for Manufacturing Innovation's shared infrastructure was envisioned to be a cost-sharing model across multiple members, seeded with federal funds but ultimately self-sustaining, with a focus on technology innovation and a strong brand identity and reputation. Because these institutes would be hubs for the nation's leading experts in individual technologies, they would be able to translate technology breakthroughs into products and businesses for the market, and form effective teams of industrial and academic experts from multiple disciplines to solve difficult problems from pre-competitive research to proprietary technology or product development.

The benefits of the National Network for Manufacturing Innovation do not end with the commercialization of new technology. The Institutes also allow for dual appointments of faculty and students in both research universities and application-oriented institutions to develop leaders familiar with research applications, new technologies and production systems. They also provide an opportunity to engage and assist small and medium-sized entrepreneurs by providing highly trained personnel to work in multiple regional innovation centers, as well as assist community colleges and universities in developing and offering courses in various manufacturing technologies so that we have a workforce ready to design, build and operate the new plants or new technology when it arrives at the "factory floor".

The Advanced Manufacturing Partnership supports the Administration's actions thus far to launch four manufacturing institutes addressing critical manufacturing technologies (Table A1 of Appendix A), with four more manufacturing institutes on the way. In support of these pilot efforts to develop the National Network for Manufacturing Innovation and in anticipation of bipartisan legislation to formally establish the program, the Advanced Manufacturing Partnership recommends the development of a shared National Network for Manufacturing Innovation Governance Structure to ensure a return on investment for manufacturing institute stakeholders.

❖ **Recommendation #5:** Create – through the National Economic Council, the Office of Science and Technology Policy, and the implementing agencies and departments – a shared National Network for Manufacturing Innovation (NNMI) governance structure that can ensure a return on investment for the NNMI's many stakeholders by including input from various agencies as well as private sector experts, organized labor and academia.

This governance structure should be established through clear, written guidelines covering both Network governance and Institute topic selection, and these guidelines should reflect multiple future scenarios for the National Network for Manufacturing Innovation: a scenario in which it is fully authorized and a scenario in which it continues to be developed through executive action. The Advanced Manufacturing Partnership proposes a governance structure that maintains autonomy for individual institute operations while creating a public-private network governing council that oversees the broader performance of the network and the sustainability of the individual institutes. For a more detailed set of principles guiding the development of the governance structure and a proposed structure that ensures all of the National Network for Manufacturing Innovation's stakeholders are adequately represented, please see Appendix 3 and related annexes.

In addition, to support the establishment of the National Network for Manufacturing Innovation, the Advanced Manufacturing Partnership engaged with leading IP experts to collect best practices on IP management to inform the practices used by the new manufacturing institutes as they are established. More detail on these best practices can be found in Appendix 3 and related annexes.

Pillar 2: Securing the Talent Pipeline

This pillar is focused on the challenges of attracting and training top U.S. talent in the varied careers within manufacturing. The Demand-Driven Workforce working team was charged to implement scalable solutions that will shift the public's perceptions of manufacturing to reflect a vibrant, creative and innovative career choice, and connect our future talent pipeline with demand-driven training that makes the best use of the expertise in community colleges and universities, meets the workforce needs of employers, and offers highly-skilled, satisfying careers to manufacturing employees.

Opportunities to accelerate U.S. manufacturing will require public-private partnerships to identify and implement the best solutions quickly and efficiently, meeting the real-time demands of manufacturers. Manufacturers and academia realize that identifying and responding to technological innovation can be expensive and time-consuming, but these opportunities are best addressed through partnerships of committed and motivated groups. With 75 percent of manufacturers being impacted negatively by a skills shortage according to one survey, the status quo will not suffice. The Advanced Manufacturing Partnership therefore offers several best-in-class models that will connect talented employees with high-paying careers, and create a robust talent pipeline that will attract manufacturers back to U.S. shores.

Shifting the Public's Misconceptions of Manufacturing

For decades, workers flocked to manufacturing careers because those jobs were viewed as stable, solid careers that provided a path to the middle class for workers at every educational level. That perception has been shaken by the job losses of past decades, and rebuilding it requires two things. First and foremost, it requires continued, real, and sustainable growth in these jobs. But it also requires convincing the public that this job growth in manufacturing is real and durable. A comprehensive, multiplatform campaign to drive home this message that the American manufacturing sector is growing again is a key component in bringing about this attitude shift. The Advanced Manufacturing Partnership reached consensus that the new image of manufacturing should be conveyed as "interesting, innovative, impactful, and most of all *increasing*", and has developed the steps required to launch a sustained, public-private media campaign, including capitalizing on National Manufacturing Day.

In order to lay the foundation of a secure, sustainable manufacturing talent pipeline, the Advanced Manufacturing Partnership recommends that manufacturing image and engagement campaigns target students at all educational levels, and primarily focus on grades K-12, school counselors, parents, and the general public in the manufacturing image campaign. Efforts should also include leveraging National Manufacturing Day as a way to engage with schools and communities. Please refer to Appendix 5 for additional information.

We recommend a layered approach to the manufacturing image campaign focused on the current and potential growth of American manufacturing to counter public perceptions of sector in decline. One-time national campaigns, while important, will not be effective without simultaneous regional and local deployment over a variety of platforms, including social media and specific activities for each age group. AMP2.0 recommends the Executive Office of the President, in concert with Department of Commerce, support a campaign that would be implemented by the stakeholders across the manufacturing community and launched using National Manufacturing Day.

National Manufacturing Day is an excellent opportunity for regional and local engagement to change public perceptions about manufacturing (traditionally held in autumn, with the next event occurring on October 3, 2014). Co-produced by Fabricators & Manufacturers' Association and the National Association of Manufacturers, the NIST MEP and other organizations, the National Manufacturing Day occurs throughout the United States and aims to address public perception of manufacturing by inviting the public in to manufacturing facilities, and makes ample use of social media to coordinate and advertise events. We recommend leveraging National Manufacturing Day as one component of this sustained image campaign, specifically as a way to engage manufacturers with local schools and communities.

AMP2.0 SC encourages its members (Appendix B) and the DOC Manufacturing Council to join forces in promoting the National Manufacturing Day, encouraging production companies all over the country to open their doors to students and the public. Further, the White House's Office of Science and Technology Policy (OSTP) can facilitate involvement of U.S. federal manufacturing facilities, such as the DOE Manufacturing Demonstration Facilities (MDFs). This expanded effort and participation from small- and medium-sized manufacturers (SMMs), large and global manufacturers with U.S.-based facilities, and "maker spaces" in and near universities will speed the improved understanding of manufacturing's image and opportunity in the United States.

- **Recommendation #6:** Launch a national campaign to change the image of manufacturing and support National Manufacturing Day's efforts to showcase real careers in today's manufacturing sector.

Connecting More Americans with Skills for Successful Careers in Manufacturing

Today, in the United States, there are currently 312,000 job openings in manufacturing (Bureau of Labor statistics), nearly half the number of jobs created in manufacturing since the end of the recession. In a 2014 study by Accenture, conducted in collaboration with The Manufacturing Institute, more than 50 percent of companies plan to increase U.S.-based production by at least five percent in the next five years and nearly a quarter are planning to grow U.S.-based manufacturing roles by almost 10 percent in the next five years. Importantly, nearly 80 percent of their manufacturing roles fell into the categories of skilled (Associate degree or equivalent 12-24 months of training and/or experience) or highly skilled (Bachelor's degree or equivalent of 36+ months or more of training and/or experience).

The work underway by AMP2.0 is critically important: it demonstrates the importance of public-private partnerships, not just increased spending, to create unique, new models that deliver demand-driven skills training at a time when the stakes have never been higher. AMP2.0 studied

studied the programs and partnerships that have proven results, and critically evaluated them for scalability and the ability to replicate them with other partners, and then widened the reach of our recommendations to tap specific subsets of the talent pool, including veterans. These models range from a broader and deeper use of accreditations and certifications, to relevant apprenticeships across a wider range of career tracks, and included using career pathway models to introduce flexibility into the training or retraining process. They all have one important goal: to connect motivated Americans with rewarding, highly skilled careers in manufacturing.

Credentialing Systems

The use of credentials and accreditation is not new in the workforce development system. Since the early 1900s, tradesmen have used a system of apprenticeships and trade school programs to develop a highly valued skill and earn certificates or accreditations after years of training and study. Professional credentials that are meaningful nationally are of tremendous value for a mobile employee, however are not always recognized or sought by employers looking to fill related positions. The use of credentials in the United States can also be sporadic and employer-dependent since not all skilled trade professions have a credential program.

Over more than a decade, dedicated efforts by organizations like the National Association of Manufacturers' Manufacturing Institute have put forth new credential programs to help address the skills gap in manufacturing. Credentialing programs, however, are limited by the extent to which employers adopt them and preferentially hire candidates with them. The goal of AMP2.0 was to find a way to increase the use of nationally portable, stackable credentialing systems through certifications and work-based learning elements, consistent with the original AMP recommendations. AMP2.0 worked closely with the Manufacturing Institute to determine best practices and to establish guidelines for credentialing and certification programs.

The best practice models and success stories that were examined by AMP2.0 indicated that credentials generally lead to higher wages, a better-trained workforce, greater labor market mobility, reduced selection costs, and higher quality employees. These initial findings indicate that credentialing systems can be an important part of the workforce of the future. By building awareness, demonstrating evidence of success, providing a national roadmap for action, and investing in certifications, we can increase the use of nationally portable and stackable credentials and better meet the needs of employers and a growing non-traditional workforce.

After creating a list of Best Practice Models (see Appendix 2), AMP2.0 identified the following common elements of credentialing systems that made them successful, including easy entry and exit points throughout a person's career, modularized certificate training that can be scheduled to meet the needs of working adults, alignment with for-credit programs leading to degrees, eligibility for education grants like Pell and WIA, and a strong link to employment or internships.

We also determined that intermediaries could organize, advocate for, and provide services to employers. In addition, intermediaries could play a role in organizing, advocating for, governing and advising workforce institutions, and engaging in research and development. Important intermediaries are state or regional manufacturing and employer associations, organized labor, workforce investment boards, community colleges, manufacturing extension partnerships, and regional economic development authorities.

AMP2.0 identified four issues with credentialing systems that if resolved could help achieve the objective of increasing nationally portable and stackable certifications: lack of awareness of certifications available and how to utilize them in both companies and educational institutions; a lack of demonstrated evidence of success of credentialing systems; a missing national roadmap for action that identified guidelines and process steps in implementing credentialing systems; and inadequate funding for implementing credentialing systems that created roadblocks for statewide systems trying to implement credentialing systems. Examining these issues in greater detail allowed us to produce a number of recommendations and project deliverables that could help resolve potential problems with certification systems.

> ❖ **Recommendation #7:** Incent private investment in the implementation of a system of nationally recognized, portable, and stackable skill certifications that employers utilize in hiring and promotion, by providing additional funds that build on investments being made through the Department of Labor and Department of Education Trade Adjustment Assistance Community College and Career Training (TAACCCT).

Apprenticeships

AMP2.0 recognized the tremendous potential that demand-driven workforce solutions could offer manufacturers, and implemented several models that will generate long-term employment opportunities. AMP2.0 identified another key mechanism for demand-driven workforce training and education that are pragmatic, scalable and sustainable: apprenticeships.

An important part of the credentialing and accreditation program for many skilled trades, apprenticeships are used to reinforce classroom lessons through hands-on learning for which trainees are paid increasing wages as their skills grow. At a time when college tuition has soared, soared, this can be an attractive career path for a motivated problem-solver. There are currently around 398,000 registered apprentices in the United States in programs run by labor unions and individual companies, according to the Labor Department. While that number sounds significant, it is actually significantly lower than in the early 2000s, when there were about 490,000 apprentices in the U.S. Compare this number to the German apprenticeship system, which trains around 1.5 million people per year.

To advance this concept, AMP2.0 has developed a best-in-class apprenticeship model with a coalition of companies and in partnership with labor market intermediaries and local colleges, and has captured the learning and best practices from these and past apprenticeship programs in a "How To" Instruction Manual, or "Playbook", targeted specifically for employers (Appendix 2 and associated annexes). The pilot with three major manufacturers and two colleges in northern California and southern Texas will prove the apprenticeship model to be a reliable, valid and repeatable process that others can implement. This model of apprenticeships entails completion of Associate Degree and Department of Labor Certification, and will be completed in September 2014. A statewide consortium of colleges, centers, and more than 25 manufacturers has also been formed in Minnesota to launch the Learn, Work, Earn model to align trainings and fill jobs. Best practices will be shared for replication. The Learn, Work, Earn model includes employer-driven competency-based apprenticeships, curricula alignment with national credentialing systems and bridging modules for veterans and other underrepresented populations. This consortium ties together all of the most important AMP2.0 concepts and recommendation in one public-private partnership that can serve as a model nation-wide. This work builds on, and should be coordinated with, the successful union apprenticeship programs.

Career Pathways

Skills credentials, accreditations and the use of apprenticeships to attain them are critical to developing the workforce for which manufacturers will compete. It is, however, important to recognize that flexibility and individuality are extremely important as new employees enter the workforce, and as mature employees retrain for the manufacturing jobs of today and tomorrow.

There is no denying that manufacturing careers have changed over the last 30 years. Many of these require specialized and unique skills, and in some cases with secondary educations of varying kinds. Community colleges stand in the gap between manufacturers who are challenged to fill openings with limited talent pools, and the local talent that may need training or retraining to fill these positions. AMP recognized the need to invest in improvements in community colleges, as well as promote engagement between community colleges and industry, universities, national labs, and K-12 programs as the most efficient and effective way in which to manage demand-driven skills training in local manufacturing ecosystems.

The analysis conducted by AMP2.0 has led to the conclusion that increasing career pathways and "dual credit" opportunities across education (K-12 schools, community colleges, and universities) will increase the number of qualified technical employees in advanced manufacturing. Career Pathways is a workforce development strategy used to transition workers from an education program to the workforce, and the goal is to increase education and job training options along certain career paths. Career pathways are a collection of programs and services intended to marry core academic competency with technical skills that are in demand. These career pathway initiatives usually consist of a partnership among community colleges, primary and secondary schools, workforce and economic development agencies, employers, labor groups and social service providers.

Efforts were therefore focused on providing job market clarity to stakeholders for the manufacturing sector and identifying various educational pathways that are linked to these in-demand careers. AMP2.0 also identified best practices in advanced manufacturing education and training from K-12 through Master's degrees.

Best-in-class curriculum for K-6 focuses on students getting satisfaction from making useful tangible objects from materials, developing and demonstrating an appreciation for the power of team-work and exhibiting discipline balanced by ingenuity. Students at this level must be trained to take calculated risks, and should be assessed not by the failure, but by successful recovery from failure, which encourages trying new ideas. At the seventh to twelfth grade level and beyond, the focus of programs was on the key attributes necessary to develop the workforce required by advanced manufacturing in sustainable, scalable models, which can include elements such as career pathways, credentialing, experiential learning, skill based certifications, blending of online and traditional program delivery, and others. Best practices from University-level programs include those with strong industry connection, such as co-op programs focused on manufacturing and professional Masters' programs. Efforts are required to assist community colleges and universities in developing and offering curricula and courses in advanced manufacturing technologies so that there is a workforce, including engineers and technologists, trained to design, build and operate the new advanced manufacturing with new technologies..

- ❖ **Recommendation #8:** Make the development of online training and accreditation programs eligible to receive federal support through federal jobs training programs.

In addition to identifying the best-in-class attributes of successful programs, AMP2.0 focused on leveraging existing efforts that are conducive to scaling. Further, AMP2.0 put these Career Pathways concepts into practice with an on-going initiative by GLOBALFOUNDRIES.

AMP2.0 has also identified efforts already underway in upstate New York to develop a scalable platform to facilitate the sharing of best practices, better connect businesses to education, people to careers and educational pathways, and in general, facilitate better connections between programs, and the needs of teachers and students. This effort is in its fourth year of development.

The pilot included a web-based, user friendly, scalable pathways portal that is generic and intended to be made available to regions across the country. The New York model can be used as an example for other regions developing such web based career pathways tools to ultimately provide a national platform (or to be used as a guide for a national platform) for building a career pathways model for advanced manufacturing education. Refer to Appendix 2 and associated annexes for more information.

Continued curation of the documents developed by AMP2.0 to address workforce training is required. These documents, described in Appendix 2 and available in associated annexes, should be updated periodically so that the best practices and relevant implementation details remain current. Manufacturing Institute, a trade society, can serve this role that benefits the wider manufacturing workforce education and training community.

❖ **Recommendation #9:** Curate the documents, toolkits and playbooks that have been created by AMP2.0 to further scale and replicate these important talent development opportunities, via the Manufacturing Institute.

Veterans

Manufacturers know that veterans have many of the work-life and jobs skills in high demand. They are committed, motivated, and highly skilled. They are disciplined leaders that are a tremendous asset to any private sector firm. Of importance to manufacturing in particular, many veterans have undertaken extensive, hands-on training with cutting edge equipment and technology over many years, resulting in high-demand skills that could be easily transferred to manufacturing positions. Careers as technicians, operators of complex equipment or craftsmen are ideal fits for our returning veterans. With more than 400,000 veterans leaving the service in the next two years, they represent a very important piece of the skills gap solution.

Transitioning from a military career to one in the private sector, however, can be a frustrating effort for our veterans in two important ways: little awareness of the career opportunities in advanced manufacturing, and difficulty equating military skills with private sector job qualifications. Employers also have difficulty understanding a military resume' and equating it with the skills they need in a new hire.

In order to alleviate this frustration and miscommunication between our veterans and employers, AMP2.0 developed practical competency based "bridging modules" for transitioning U.S. Veterans focused on private sector manufacturing skills certifications. AMP2.0 created two tools to aid veterans, employers and colleges in moving this skilled, dedicated piece of the U.S. talent pool into private sector careers. First, AMP2.0 built an inventory of veteran resources, including key tools and applicable links that catalog and summarize the resources for Veteran use (Appendix 2 and associated annex). AMP2.0 also wrote a practical guide for Veterans, Employers and Academic Institutions on Transitioning Veterans to non-military advanced manufacturing roles (Appendix 2 and associated annex). In addition to these two important tools, the AMP2.0 provided recommendations for skills translators and veterans' skill badging that enables transitioning military personnel to translate military training and experience to civilian careers.

Refer to Appendix 2 and associated Annexes 11-24 for detailed information on these workforce training-related recommendations and actions.

Pillar 3: Improving the Business Climate

AMP recognized that there were policy choices that could be made to affect real change in the manufacturing sector. While many of the initial recommendations were focused on macroeconomic policy, AMP identified early on that there were also specific and targeted policy interventions that could be made to foster the scale-up process in small- and mid-sized manufacturers, both start-ups and established enterprises.

"Scale-up" can be defined as the translation of an innovation into a market. There are significant technical and market risks faced by new manufacturing technologies during scale-up. The path to successful commercialization requires that technologies function well at large scale and that markets develop to accept products produced at scale. It is a time when supply chains must be developed, demand created and capital deployed. To compete globally and be a leader in innovation, the United States must significantly improve its ability to translate innovation into practical production.

There are three requirements to achieving commercial scale with promising advanced manufacturing technologies: (1) networked supply chains, (2) rapid diffusion of technology through the networked supply chains, and (3) access to capital. Barriers to achieving scale include the impeded flow of technical or market insights, supply network relationships, and funding. Small- and medium-size manufacturers (SMMs) in the United States are particularly susceptible to information, relationship, and finance barriers. U.S.-based manufacturers of all sizes experience barriers to scaling-up new manufacturing innovations, due to financial risk and the capital-intensive nature of production at market scales. Further complicating the scale-up process, support for manufacturing is largely regional and varies substantially across regions within the United States.

In addition to its efforts this year to further develop and implement its recommendations on critical emerging technology areas, securing the talent-pipeline, and creating a supportive environment for innovative manufacturing businesses, the Advanced Manufacturing Partnership reiterates the recommendations on tax and regulatory policy advanced in its original report. The Advanced Manufacturing Partnership applauds the Administration's efforts through regulatory look-backs to reduce regulatory burdens for manufacturers and its proposals on business tax reform to create a more competitive environment for manufacturing. The Advanced Manufacturing Partnership emphasizes the importance of passing fundamental business tax reform while deepening those incentives designed to encourage the long-term establishment of capital-intensive and space-intensive manufacturing operations as part of coordinated national strategies to become manufacturing leaders in a specific sector.

Please refer to Appendix 4 and associated Annex 26 for detailed information on the recommendations and actions in this section.

Expanding and Exchanging Intermediary Solutions for Manufacturing

Small and medium-sized manufacturers can adopt and leverage manufacturing technologies more readily when improved access to information is provided via "intermediary services" – established organizations that understand and facilitate access to a wide range of information needed for technology commercialization and scale up, including technical expertise, supply chain partners, financing options, and government programs. Several types of private sector, public sector, and non-profit organizations currently provide some of these services (see annex). Successful intermediaries share common characteristics (see list in Appendix 4), and will focus on a regional economic ecosystem; will scope specific technologies or industries; and will provide information, technical assistance, and other services to a network of firms rather than a single firm. The Hollings Manufacturing Extension Partnership (MEP) Program is one example of a public sector intermediary that meets some of the Design Characteristics itemized below and could be further enhanced or configured to meet the key design criteria. However, many other successful models of providers exist including NorTech (OH), the Rochester Regional Photonics Cluster (NY), The Great Lakes Wind Network, SF Made (CA), Maker's Row (NY/NJ), and Life Science Alley (MN).

Small and medium-sized manufacturers make scale-up investment choices and attain financing based on their understanding of a given business opportunity, as do their larger peers. However,

However, small firms lack the internal capacity to ascertain the potential demand generated by new technologies and understand how to enter those highly dynamic markets profitably. For these reasons, affordable and accurate market and technical insight is critical for small and medium sized manufacturers to develop an entry strategy, and mobilize resources to adopt new technologies for processes, materials, and new products.

AMP2.0 identified bottlenecks in information flow and key information types needed by small and medium manufacturers to develop a business plan for scale-up, and launched a pilot project running from May to August 2014 to illustrate an approach to provide technical and market insight, termed the Printed Electronics Pilot Project. This pilot examined a specific manufactured product (Printed Electronics), U.S. geographic region (Midwest), and established small and medium manufacturer (Mac Arthur Corporation). Results of the pilot indicate that the following aspects would be critical to a successful process to deliver technical and market insight to small and medium sized manufacturers:

- A regional, supply network hub for technology and industry expertise.
- Knowledge assets including Technology Readiness Level / Manufacturing Readiness Level (TRL/MRL) hurdles, supply network maps, capital and capability requirements for entry, and lessons learned.
- Attract supply network participants, especially SMMs and strategic industry partners, via workshops.
- Market entry support to small and medium sized manufacturers, including use of enhanced intermediary solutions such as the Manufacturing Extension Partnership centers or the National Network for Manufacturing Innovation Institutes.

The Great Lakes Wind Network (GLWN.org) and The New York Battery & Energy Storage Technology Consortium (ny-best.org) are examples of intermediaries creating market insight at scale and utilizing such information to create the supply network. These models provide market and technical insight leading to efficient and rapid supply network creation. Additionally, as discussed further in Appendix 1, for certain manufacturing technologies such as digital manufacturing, large scale technology incorporation can only emerge if trusted exchange of information and digital data occurs among diverse supply chain entities. For this purpose, trusted agent accreditation/certification by intermediary organizations will be highly beneficial.

- **Recommendation #10:** Leverage and coordinate existing federal, state, industry group and private intermediary organizations to improve information flow about technologies, markets and supply chains to small and medium-sized manufacturers.

Increasing Capital Access For Established and Start-up Firms

Emerging advanced manufacturing technology companies often are not compelling investments for capital markets due to technology risk, market adoption risk, long lead times to significant revenue and significant capital requirements. Reducing capital requirements is one way that government has encouraged the scale-up of early-stage high-technology manufacturers through grants, loan guarantees and tax deferrals. However, two modes of capital access assistance have historically provided little benefit to small and medium sized manufacturers at production scale-up stages: venture capital funding and federal tax deferrals. Although they are now returning to more traditional levels, AMP2.0 found that, in general, reporting rules enacted in the United States have adversely impacted the initial public offering (IPO) market that is part of a venture capital investment strategy, impeding the ability of small companies to explore traditional equity markets for capital. Federal tax deferrals also provide manufacturing start-ups no benefit, because they are rarely initially profitable and, therefore, are not subject to an immediate tax liability. There are, however, many local and regional incentives that exist to support manufacturing in general.

Frequently overlooked are the other means through which investments to scale-up production can be made more attractive to capital markets such as through demand creation (reducing market adoption risk), providing access to technical expertise (reducing technical risk), and through reducing development time (shortening investment maturities).

AMP2.0 held multi-city workshops to identify mechanisms beyond grants, loan guarantees and tax deferrals that make investment in manufacturing more attractive to capital markets. These workshops engaged leaders from a broad array of bank, venture, and other capital access mechanisms to share ideas and best practices in manufacturing investment.

The following strategic approaches can improve capital access for both established small and medium sized manufacturers and manufacturing start-ups looking to bring capital-intensive technologies to commercial scale production:

- *Launch a Public-Private Scale-Up Investment Fund for First At-Scale Production Facilities.* By offering low-cost loans to private-sector investors in "first-of-a-kind" production facilities a public-private Scale-Up Fund could incentivize additional investment in first of a kind production facilities, ensuring that technologies invented in the United States can be made in the United States. The fund would award loans to investment funds or investor consortia in an equivalent amount to half the cost of the project being financed, and support investments of at least $40 million, to address investments at the scale where access to finance becomes truly challenging. This mechanism can be used, for example, to incentivize a U.S. equipment supply base for nascent manufacturing technologies such as additive manufacturing. Given the need to maintain a diverse portfolio of investments and the scale of manufacturing projects (easily $40 million to more than $150 million), the public-private investment fund would need to be able to provide $5 to 10 billion in capital over time, split across private funding and public loan guarantees (where generally $1 of public funds can create the equivalent of $10 of guarantees).

- *Facilitate Connections among Corporate Strategic Partners, the Federal Government, and Small and Medium Sized Manufacturers* – Strategic partners can play a critical role in deployment of advanced manufacturing technologies and in small and medium sized manufacturers' success. The government should invest in, and be a customer of, a private-sector web-based platform that creates a portal for strategic partners to learn more about small and medium-sized manufacturers, their technologies and their capabilities. In addition, this portal could be used to advertise federal program funding that can support the scale-up of manufacturing technologies. An intermediary accreditation/certification mechanism will also accelerate adoption of digital design and manufacturing.

- *Use Tax Incentives to Foster Investment in Manufacturing and Partnerships in the Manufacturing Ecosystem.* While tax credits rarely directly benefit small to mid-sized manufacturers, tax credits that encourage investments in manufacturing start-ups and small to mid-sized manufacturers can attract additional capital from strategic partners. One such tax incentive would be created by continuing/modifying the New Markets Tax Credit Program (NMTC) and expanding its scope to build the manufacturing ecosystem through support of new and expansion projects in the form of a Manufacturing Renaissance Tax Credit. In addition, Congress should extend the Research and Development Tax Credit to allow for early stage domestic testing of commercial viability including investments in depreciable property (RD&D – Deployment).

- *Leverage Government Procurement and Demand for Small and Medium-Sized Manufacturers' Technologies.* Existing federal programs and authorities can make scale-up investments for nationally strategic manufacturing technologies, in line with a national technology strategy as proposed by AMP2.0 elsewhere in this report. The preannouncement of intended scale up investments from federal programs can help reduce the market risk associated with bringing new technologies to commercial scale by ensuring a base load of demand. The Department of Defense (DOD) and the Department of Energy (DOE), in particular, have experience with manufacturing scale-up investments from the DOD's recent efforts on energy efficiency and the DOE's investments in solar energy that could be leveraged.

  ❖ **Recommendation #11:** Reduce the risk associated with scale-up of advanced manufacturing by improving access to capital through the creation of a public-private scale-up investment fund; the improvement in information flow between strategic partners, government and manufacturers; and the use of tax incentives to foster manufacturing investments.

III. Conclusions

ACCELERATING ADVANCED MANUFACTURING REQUIRES SUSTAINED ACTION

The Advanced Manufacturing Partnership 2.0 Steering Committee has implemented action plans for a prioritized subset of the recommendations made by AMP around three critical pillars:

- Enabling Innovation
- Securing the Talent Pipeline
- Improving the Business Climate

The prioritized technology strategies, action plans, pilots, playbooks, and subsequent recommendations aim to accelerate innovation in the U.S. manufacturing in a way that ensures our global competitiveness and grows a robust domestic manufacturing base. In 12 months, we have focused our efforts on positioning U.S. manufacturing to lead the world in the new disruptive advanced manufacturing technologies and to capitalize on the inherent strengths of the United States' innovation economy.

Over the course of this year, AMP2.0 also identified several opportunities outside the original scope of AMP2.0. The AMP2.0 SC notes that there remain regulatory, tax and other policy issues that may surmount technological advantages. Manufacturers note that regulatory uncertainty impacts their ability to both adopt new manufacturing technology innovations and to continue U.S.-based operations. Second, AMP2.0 encourages the further development of tax policies that are designed to encourage long-term establishment of capital-intensive and space-intensive manufacturing operations as part of coordinated national strategies to become manufacturing leaders in a specific sector.

Accelerating manufacturing innovation requires sustained momentum of the actions begun by AMP2.0. The United States can and will lead the world in advanced manufacturing if we commit to doing so.

Appendix A: AMP Recommendations and Implementation Status

Below we provide the original list of AMP recommendations to the President, conveyed in the PCAST AMP report in 2012. AMP2.0 was not tasked to review the whole-government progress against these recommendations, but provides this table as a brief reference to the context of AMP2.0. The AMP Steering Committee recommendations to the Administration were well-received, identifying opportunities not only for government – for both the executive and legislative branches – but also for private sector industry and academia. The Administration has made substantial progress against the key recommendations, across all three 'pillars' with significant new initiatives under 'Enabling Innovation.' The White House and Commerce have been driving implementation across Federal agencies with immediate focus on executive actions and agency budgets. This includes a major new initiative announced by President Obama to launch the National Network for Manufacturing Innovation, as Congress takes action on bipartisan, bicameral legislation. We then provide a summary of actions to date on these recommendations, with a view of additional recommended actions for implementation.

AMP Recommendation (2012)	*Status Update (2014)*
1. **Establish a National Advanced Manufacturing Strategy to Identify and Prioritize Investment in Cross-Cutting Technologies**	The PCAST reports served as the private sector voice on the national strategic plan, and the Administration published the NSTC report "A National Strategic Plan for Advanced Manufacturing in February 2012. Additionally the administration formed the Advanced Manufacturing National Program Office (AMNPO) as an interagency team to coordinate and collaborate on crosscutting initiatives – including the NNMI. Further coordination responsibilities would require legislative action.

2. **Increase R&D Funding in Top Cross-Cutting Technologies**	The Administration proposed $2.2 billion in advanced manufacturing R&D in FY13 Budget, a nearly 20% increase over the prior year ***In Process / Planned Activities:*** Build process to incorporate regular input from industry, labor and academia
3. **Establish a National Network of Manufacturing Innovation Institutes**	Launched a pilot institute on additive manufacturing in Youngstown, Ohio; led by a consortium of now over 100 companies, universities, and other organizations. Engaged with stakeholders through five regional workshops on NNMI, and published the NNMI design framework in January 2013. Additional public input gathered on Intellectual Property and Institute performance measures. Launched three additional pilot manufacturing innovation institutes – in Raleigh, NC, in Detroit, MI, and in Chicago, IL – with four more manufacturing innovation institutes on their way through a competitive selection process. **In Process / Planned Activities:** Working with Congress to develop and pass legislation to launch the NNMI NNMI preparedness plan and team established.

4.	**Empower Enhanced Industry/University Collaboration in Advanced Manufacturing Research**	IRS treatment of industry-funded R&D and technology development in non-profit university facilities limits industry-university collaborations; this treatment has not changed.
5.	**Foster a More Robust Environment for Commercialization of Advanced Manufacturing Technologies**	Created $1 billion Small Business Investment Company "early stage" innovation fund to support advanced manufacturing and other emerging technologies
		Signed an Executive Order requiring all Federal research agencies to bolster efforts that support commercialization of technology (October 2011)
		Passed most significant patent reform in 50 years with bipartisan America Invests Act, which includes a fast track option for processing within 12 months
		In Process / Planned Activities: Developing partnership with the Association of University Technology Managers to enhance and measure startup and licensing activity

6.	**Establish a National Advanced Manufacturing Portal**	Revamped and re-launched the manufacturing.gov web-site
		Launched BusinessUSA, a new online platform to facilitate access to services and information businesses need across the federal government
		Created the MEP-run National Innovation Marketplace, a portal that connects U.S. manufacturers to over $2 billion in technology buying opportunities
		In Process / Planned Activities: Launching of manufacturing.data.gov, including a searchable database of research activities, as part of a larger 'Open Manufacturing Data Initiative'
		Based on the results of the pilot-tested National Innovation Marketplace (NIM), MEP launched a competition for a set of regional business-to-business network pilot projects to build on the lessons learned from the NIM.
7.	**Correct Public Misconceptions About Manufacturing**	This has been partially implemented through 2013-2014 public events, such as the July 2014 Maker Faire in Washington, D.C. and more regionally-distributed events on National Manufacturing Day. No advertising campaign has been funded or launched as of this report finalization.
8.	**Tap the Talent Pool of Returning Veterans:**	Established the Military Credentialing and Licensing Task Force at the Dept. of Defense to enable service members in the Army, Navy, Marines, and the Air Force to get industry-recognized credentials for their skills, starting with the 126,000

	active service members who receive elements of manufacturing training
	Created the <u>Veterans Retraining Assistance Program</u> to offer up to 12 months of training to unemployed veterans
	Sought and secured commitments from private manufacturing employers to hire returning veterans
9. **Invest in Community College Level Education**	NSF's Advanced Technology Education (ATE) program continues to address manufacturing skills; Department of Labor TAACCCT program also partially addresses these issues.
10. **Develop Partnerships to Provide Skills Certifications and Accreditation**	Supported 5-year goal of 500,000 workers receiving industry-recognized credentials through a partnership between Skills for America's Future and NAM, supported by the Manufacturing Extension Partnership
	Launched a Department of Labor manufacturing credentialing database
11. **Enhance Advanced Manufacturing University Programs**	Renewed emphasis on manufacturing in NSF Advanced Technological Education (ATE) program
12. **Launch National Manufacturing Fellowships & Internships**	Increased funding requested for NSF Graduate Research Fellowships (GRF), including about $250 million in fields related to manufacturing
	Proposed NIST scholarship and fellowship program in advanced manufacturing in the FY13 Budget

13. Enact Tax Reform	Not enacted to date.
14. Streamline Regulatory Policy	Launched regulatory 'lookback' across more than two dozen agencies and 19 independent agencies to remove outdated, inefficient, or redundant regulations – 580 reform proposals have been submitted and we have already acted on 100
15. Improve Trade Policy	Established goal of doubling exports over five years to support <u>2 million export-related jobs</u>, with over 1.2 million export-supported jobs created so far
	Signed into law Free Trade Agreements with Colombia, Panama, and Korea (Fall 2011)
	Created new trade enforcement unit, the International Trade Enforcement Center (ITEC), to challenge illegal trade actions of other countries
	Doubled rate of trade cases against China compared to previous administration. Significant cases including (i) rare earth WTO case against China and (ii) WTO case on Chinese subsidies for auto and auto parts manufacturing

16. **Update Energy Policy** — DOE has an ongoing significant effort to bring efficiency and conservation to the manufacturing sector through its Energy Efficiency and Renewable Energy office; DOE has announced an MII in Clean Energy Manufacturing for power electronics and has another active solicitation for fiber reinforced polymer composites.

Table A1. Manufacturing Innovation Institutes (MIIs) announced by August 2014.

Manufacturing Innovation Institute Name	Primary Federal Funding, Solicitation, & Management Entity	Announced Date	Total Federal Investment (over 5 years, in USD)
America Makes National Additive Manufacturing Innovation Institute	DOD-Air Force	Aug 2012	$50M
Power America Next-Generation Power Electronics National Manufacturing Innovation Institute	DOE-Energy Efficiency and Renewable Energy	Jan 2014	$70M
Digital Manufacturing and Design Innovation Institute	DOD-Army	Feb 2014	$70M
American Lightweight Materials Manufacturing Innovation Institute Lightweight and Modern Metals Manufacturing Innovation Institute	DOD-Navy	Feb 2014	$70M
Clean Energy Manufacturing Innovation Institute for Composites Materials and Structures	DOE-Energy Efficiency and Renewable Energy	TBD [review of proposals in process]	$70M

Appendix B: AMP2.0 Membership and Participation

Operating under the framework of PCAST, the Advanced Manufacturing Partnership 2.0 is led by a Steering Committee co-chaired by Andrew Liveris, President, Chairman and Chief Executive Officer of the Dow Chemical Company, and Professor Rafael Reif, President of the Massachusetts Institute of Technology. Working closely with White House's National Economic Council and PCAST, the AMP2.0 Steering Committee brought together a broad cross-section of U.S. manufacturers and colleges and universities.

The AMP2.0 Steering Committee is comprised of the following members:

- Wes Bush, Chairman, CEO and President, Northrop Grumman
- David Cote, Chairman and CEO, Honeywell
- Nicholas Dirks, Chancellor, University of California, Berkeley
- Kenneth Ender, President, Harper College
- Leo Gerard, International President, United Steelworkers
- Shirley Ann Jackson, President, Rensselaer Polytechnic Institute
- Eric Kelly, President and CEO, Overland Storage, INC
- Klaus Kleinfeld, Chairman and CEO, Alcoa, INC
- Andrew Liveris, President, Chairman and CEO, The Dow Chemical Company
- Ajit Manocha, Senior Advisor, GLOBALFOUNDRIES
- Douglas Oberhelman, Chairman and CEO, Caterpillar, INC
- Annette Parker, President, South Central College
- G.P. "Bud" Peterson, President, Georgia Institute of Technology
- Luis Proenza, President Emeritus, The University of Akron
- Rafael Reif, President, Massachusetts Institute of Technology
- Mark Schlissel, President, University of Michigan
- Eric Spiegel, President and CEO, Siemens Corporation
- Mike Splinter, Executive Chairman of the Board, Applied Materials, INC
- Christie Wong Barrett, CEO, Mac Arthur Corporation

In addition to the contributions of the Steering Committee members, there were numerous other individuals whose time and commitment to AMP2.0 should not go unrecognized. The following individuals were either AMP2.0 working team members from academia, industry, and labor or federal agency and administrative support.

Table A2. AMP2.0 Participants

Deborah Altenburg, Rensselaer Polytechnic Institute	Pooja Anand, Siemens Corporation	Brian Anthony, Massachusetts Institute of Technology
Michael Barriere, Alcoa	Greg Bashore, Alcoa	Randy Belote, Northrop Grumman
Gisele Bennett, Georgia Institute of Technology	Stacey Bernards, Honeywell	Wieslaw Binienda, The University of Akron
William Bonvillian, Massachusetts Institute of Technology	David Bridges, Georgia Institute of Technology	Tina Brown, Overland Storage INC
Mike Brown, Siemens Corporation	Travis Bullard, GLOBALFOUNDRIES	Rolf Butters, Advanced Manufacturing National Program Office
Jennifer Clark, Georgia Institute of Technology	Mary Sue Coleman, University of Michigan	Maria Coons, Harper College
Hope Cotner, Center for Occupational Research & Development	Marsha Danielson, South Central College	Brian Davis, The University of Akron
Oliver de Weck, Massachusetts Institute of Technology	Kimberly Denley, Siemens Corporation	Mitchell Dibbs, The Dow Chemical Company
Nicholas Dirks, University of California, Berkeley	Jonathan Dordick, Rensselaer Polytechnic Institute	David Dornfeld, University of California, Berkeley
Craig Dory, Rensselaer Polytechnic Institute	Joseph Dragone, Ballston Spa Central School District	Barb Embacher, South Central College
Kenneth Ender, Harper College	Michael Engelhardt, Independent	Joseph Ensor, Northrop Grumman
Mark Esherick, Siemens Corporation	Sergio Felicelli, The University of Akron	Karen Fite, Georgia Institute of Technology
Randy Gast, Overland Storage	Frank Gayle, Advanced Manufacturing National Program Office	Christopher Gopal, Independent
Charles Grindstaff, Siemens Corporation	Dave Gross, GLOBALFOUNDRIES	Craig Habeger, Caterpillar INC
Rod Heiple, Alcoa INC	Carrie Houtman, The Dow Chemical Company	Jack Hu, University of Michigan
Scott Hudson, Alcoa INC	Shank Iyer, Honeywell	Michael Jackson, Department of Commerce

Matt Jensen, Alcoa INC	Camille Johnston, Siemens Corporation	Mark Jones, The Dow Chemical Company
AJ Jorgenson, Manufacturing Institute	Tom Kammer, South Central College	John Kania, Applied Materials, INC
Ray Kilmer, Alcoa INC	Robert Knotts, Georgia Institute of Technology	Brian Knutson, South Central College
Kristina Ko, University of Michigan	Kevin Kolevar, The Dow Chemical Company	Tom Kurfess, Georgia Institute of Technology
Doug Laven, South Central College	Lance Lavergne, Alcoa INC	Douglas Lawton, Northrop Grumman
Lori Lecker, Alcoa INC	Philip Lippel, Massachusetts Institute of Technology	Brian Lombardozzi, United Steelworkers
Helmuth Ludwig, Siemens Corporation	Ajay Mahajan, The University of Akron	Erin Makarius, The University of Akron
Marissa McCluney, Siemens Corporation	Michael McGrath, Analytical Services, INC	Jennifer McNelly, Manufacturing Institute
Shreyes Melkote, Georgia Institute of Technology	Krishna Mikklineni, Honeywell	Jason Miller, National Economic Council
Lauren Minisci, Siemens Corporation	Mike Molnar, NIST and Advanced Manufacturing National Program Office	Padraig Moloney, Lockheed Martin Company
Liz Moress, University of California, Berkeley	Gopal Nadkarni, The University of Akron	Michael Nobel, Caterpillar INC
Michelle O'Neill, Alcoa INC	Esra Ozner, Alcoa INC	Mike Panigel, Siemens Corporation
Nag Patibandla, Applied Materials, INC	Scott Paul, United Steelworkers	Brian Paul, Advanced Manufacturing National Program Office
Ashley Predith, PCAST	Doug Ramsey, Alcoa INC	Jessica Raynor, National Economic Council
Liz Reynolds, Massachusetts Institute of Technology	Bill Ritsch, Georgia Institute of Technology	Ignacio Ros, Siemens Corporation
Don Rosenfield, Massachusetts Institute of Technology	Mike Russo, GLOBALFOUNDRIES	Tariq Samad, Honeywell
Shivakumar Sastry, The University of Akron	Mike Sayen, Siemens Corporation	Natalie Schilling, Alcoa INC

Richard Schmaley, Northrop Grumman	Martin Schmidt, Massachusetts Institute of Technology	Ravi Shanker, The Dow Chemical Company
Albert Shih, University of Michigan	Rajiv Singh, Honeywell	Lisa Skaggs, The Dow Chemical Company
Erin Sparks, Department of Commerce	Mahesh Srinivasan, The University of Akron	Marianne Stanke, Motorola Solutions
Rebecca Taylor, National Center for Manufacturing Sciences	Rainer Theisen, Siemens Corporation	Paul Towne, Honeywell
Christopher Traci, United Steelworkers	Marion Usselman, Georgia Institute of Technology	Krystyn Van Vliet, Massachusetts Institute of Technology
Gina Vassallo, Caterpillar	Ben Vickery, Advanced Manufacturing National Program Office	Christopher Voight, Massachusetts Institute of Technology
Ben Wang, Georgia Institute of Technology	Wayne Watkins, The University of Akron	John Wen, Rensselaer Polytechnic Institute
Michael Wessel, United Steelworkers	Gloria Wiens, Advanced Manufacturing National Program Office	Paul Witt, The Dow Chemical Company
Jeff Zawisza, The Dow Chemical Company	Chuck Zhang, Georgia Institute of Technology	

Appendix C: U.S. Regional Meetings on Advanced Manufacturing during the Advanced Manufacturing Partnership 2.0

Each regional meeting listed below was hosted by institutions that were participating members of the AMP2.0 Steering Committee and/or Working Teams. The agenda for each meeting was specific to that region, focused chiefly on regional planning for manufacturing innovation with the state and its neighboring states. See URLs listed below for complete agendas and speakers. Each meeting was advertised widely by AMP2.0 and the AMNPO. The audience for each meeting typically exceeded 200 attendees from regional industry, academic institutions, local and state governments, and federal agencies.

Each regional meeting's organizational efforts and costs were volunteered by the hosting institution(s). No formal summary of the meeting was requested or required by AMP2.0, and generally these events served the purposes of obtaining:

- Regional discussion and feedback to AMP2.0 process and recommendations;
- Showcasing of regional manufacturing strengths; and
- Spurring regional planning among local industry, academia, and government.

Table A3. AMP2.0 regional meetings

DATE	LOCATION	HOSTING INSTITUTION(S)	MEETING AGENDA URL
February 3, 2014	Atlanta, GA	Georgia Institute of Technology	http://www.advancedmanufacturingatech.edu
April 2, 2014	Akron, OH	University of Akron, United Steelworkers	https://www.uakron.edu/amp/
April 24, 2014	Troy, NY	Rensselaer Polytechnic Institute and Global Foundries	http://amp.rpi.edu
May 16, 2014	Cambridge, MA	Massachusetts Institute of Technology and the Commonwealth of Massachusetts	http://ilp.mit.edu/conference.jsp?confid=111
June 9, 2014	Detroit, MI	University of Michigan and Northrup Grumman (Lead-in event to "Big M" manufacturing meeting, June 9-12)	http://www.sme.org/amp/

Appendix 1: Transformative Manufacturing Technologies

Background

AMP2.0's Transformative Manufacturing Technology Working Team was charged to analyze and recommend actions for a subset of 11 manufacturing technology areas that were identified by the Advanced Manufacturing Partnership in its first report. The term "manufacturing technology area" or MTA is broad, and is intended to convey a set of related technical capabilities needed to synthesize and assemble products at industrial scales. Indeed, it is the sustained innovation of those technologies that is at the heart of U.S. manufacturing resurgence, and thus the detailed findings and recommendations in Appendix 1 are considered a key output of AMP2.0's report.

This appendix summarizes the process by which AMP2.0 prioritized a wide range of important MTAs, and then for three MTAs identified technology and implementation gaps that were addressed via specific suggestions; recommendations appear in the main AMP2.0 report. Those self-contained letter reports are provided in Annexes 1-10. The goal of this effort was to collectively identify actions that achieved impact in short (3 year) and long (>20 year) time scales across a variety of public and private technology investments.

Process & Deliverables

This team met via biweekly teleconferences, and in person at AMP2.0 regional meetings in GA, OH, and MA. Three MTAs were selected for detailed analysis of gaps and recommended actions, through a process outlined in Appendix 1, and a subteam was formed to develop each MTA strategy. These three prioritized MTAs included: Advanced Sensing, Controls and Platforms for Manufacturing (ASCPM); Visualization, Information, and Digital Manufacturing (VIDM); and Advanced Materials Manufacturing (AMM).

Incidentally, this triplet of MTAs provided a range of technology maturity and of current federal investment. ASCPM describes the infrastructure of hardware and software for smart and safe manufacturing. VIDM describes the virtual design, testing, and integration of manufactured components, and was related to what the recently awarded pilot institute on Digital Manufacturing and Design Innovation would address. Finally, AMM describes a broad set of challenges in U.S.-based design, synthesis, and processing of both mature and novel materials that we further subdivided into timely case studies: structural composites, biotherapeutic manufacturing, and critical materials reprocessing In fact, a new NNMI pilot institute solicitation in advanced structural composites was announced after AMP2.0's analysis of that topic was already underway.

Expert input was solicited from non-AMP participants for each MTA, and facilitated through teleconference discussion of draft "letter reports" for each MTA that AMP2.0 then finalized. Deliverables include letter reports in Annexes 1-10, each describing strategies to achieve U.S. strength in these MTAs over short and long time scales. Those annexes are highly detailed with analysis of the current landscape, vision, technical and implementation challenges, and analysis of investment and partnership opportunities in each MTA.

Rather than restate those MTA-specific findings here, we provide three broad findings gleaned from our intensive analysis of multiple and disparate MTAs. These findings and associated recommendations in the main AMP2.0 report address challenges in U.S. manufacturing innovation that cut across specific technologies (e.g., digital and materials manufacturing) and across specific sectors (e.g., aerospace and biotech).

Prioritization Process: Selection of Manufacturing Technology Areas (MTAs)

The United States currently lacks a national technology strategy that identifies a targeted list of high priority MTAs to guide public and private investment across the TRL/MRL spectrum. AMP provided a list of 11 MTAs, at varying levels of specificity and maturity. Each agency has a defined mission into which such manufacturing technology areas must fit, and thus coordination and identification of MTAs of high national interest is not readily addressed by this process. Thus, AMP2.0 sought to create and model a process by which the nation could identify which MTAs should be prioritized for coordination and funding, given finite public and private resources. We do not believe this should be a process to pick winners, in terms of any specific product or company or even means to meet manufacturing technology gaps. Rather, this is a process by which we chose among many important and interesting MTAs because our time and resources to create detailed strategies was also finite.

MTA Prioritization Criteria: AMP2.0 identified four criteria by which we assessed priority of a given MTA:

1. <u>Industry or market pull:</u> Does there exist a current "pull" or demand for this MTA by industry? If industry is not yet adopting this MTA, is there a strong perceived pull by the market or consumers?

2. <u>Cross-cutting</u>: Does this MTA cut across many sectors (automotive, aerospace, biotech, infrastructure), and across multiple sizes of manufacturers in the supply chain network?

3. <u>National or economic security</u>: Does failure to have U.S. competence or dominance in this MTA pose a threat to national security or to economic security? Does lack of U.S. competence severely disadvantage the U.S. competitiveness position of the supply network?
4. <u>Leveraging U.S. strengths</u>: Does this MTA leverage an already available workforce and education system, unique infrastructure, or policies?

Voting, Weighting, and Selection: Each of the above prioritization criteria was rated on a scale of 1 (low) to 4 (high), and assigned equal weight. A draft poll was taken to assess feasibility and weighting of criteria. Each AMP2.0 partner in WT-1 voted (i.e., one vote per institution), and results were compiled to rank the 11 MTAs from AMP; the scores in the final tally conducted by online voting were unweighted by either the criterion or the voting partner (e.g., large and small companies received an equal vote). The top four MTAs, listed here according to the original AMP names, were thus identified as Advanced Materials Manufacturing; Advanced Sensing, Measurement & Process Control; Visualization, Information, and Digital Manufacturing; and Industrial Robotics.

Scoping of MTAs for final prioritization: We subdivided the WT into four subteams, each researching and writing a "Framework Document" on one of the above MTAs over the course of one month. Outside experts were contacted, and these documents were all designed to address the current landscape, key participants, vision for the future, technical gaps, and draft recommendations. The WT met to review these framework documents, and decided that the top 3 MTAs should each be developed in further detail over the next four months; the fourth MTA of industrial robotics was deemed important but already well scoped in terms of the Robotics Initiative and other efforts, and was not developed further by AMP2.0. We then went on to conduct detailed analysis of AMM, ASCPM, and VIDM as detailed in Annexes 1-9.

Suggestions across all Manufacturing Technology Areas for Sustainable Progress in U.S. Manufacturing Innovation

A. **Formalize Sustainable Strategy Development and Governance**: We suggest for consideration several follow-on steps to help ensure the continuing effectiveness of the government's work on advanced manufacturing, and to maintain industry engagement. Below we propose continued, national-scale development of advanced manufacturing technology strategies; coordination of agency research & development (R&D) in advanced manufacturing; and an advisory process that maintains channels expert input. AMP2.0 suggests a process that employs three mechanisms.

1. **Advanced Manufacturing Technology Strategies:** AMP2.0 developed a process to identify among many important MTAs those of high priority and impact, and then to create strategies that would address technological and implementation gaps for each prioritized MTA. The process and deliverables outlined below have validated the concept of such collaborative strategies formed among industry, academia, and labor representatives, including input from many experts beyond the AMP2.0 member organizations. These and additional strategies could contribute to informed technology development in industry and in government supported R&D and applied programs. Based on the AMP2.0 experience, it will be important that agency experts who can represent the involved agencies' mission and support mechanisms be full participants in the development of the strategies, and that experts from industry and academia can provide detailed input on technological and implementation challenges. Not all strategies need to be undertaken at the same time; one or two initial pilot strategies could be undertaken to develop best practices for the process at the national level. Prioritization and investment in cross-cutting technology strategies, which should be ongoing, can play a dynamic role in enabling advanced manufacturing along the supply network and in motivating public-private partnerships across sectors (e.g., automotive, aerospace, biotech).

    AMP2.0's process demonstrated that the following considerations critically impact technology strategy development and deployment. First, the technology vision must embed both short-term (3 to 7 years) and long-term (7 to 20 years) technology and manufacturing readiness levels. Second, equal consideration must be given to the technology-push perspectives of academia and certain government agencies, and to the market-pull perspectives of industry and other government agencies. Third, open Requests for Proposals (such as NIST's AMTech) can identify qualified teams to research and scope gaps and challenges without proposing how best to solve them; and existing or proposed federal investments such as the NNMI should immediately benefit from such analysis to augment planning. Fourth, all relevant government agencies should be provided early input in strategy formulation to facilitate alignment of national goals with agency mission requirements. Fifth, existing government mechanisms should be known, documented, and realigned as possible to implement strategies with respect to authorization and appropriation legislation. Sixth, accountability to a larger advisory committee and interagency process promotes coherent progress and deliverables.

2. **Establishment of an Advanced Manufacturing Advisory Consortium:** Based on AMP2.0 progress and outcomes, we believe it is important to ensure continuing industry, organized labor and university expert advice and input to the federal government agencies working on advanced manufacturing. To facilitate interagency linkages and convey the desired experience and time commitment of its members, we suggest that an Advanced Manufacturing Advisory Consortium (AMAC) be created. Its major task would be in developing, advising, and updating on MTAs, as well as associated workforce education and policy issues as requested. The AMAC should comprise members working in industry and universities with expertise and experience in the range of critical advanced manufacturing fields. AMAC should serve as a sustained channel for detailed, coordinated input on nascent opportunities and priorities in manufacturing that can shape U.S. spending priorities from the low to high TRL/MRLs. AMAC should be tasked with manufacturing technology analyses, similar to the process demonstrated by AMP2.0 for three Manufacturing Technology Areas, and with providing recommendations for action against the periodically updated national strategy. This consortium should meet at least annually and interface regularly with the government leadership described in Recommendation 1 to provide feedback or partnership in strategy implementation.

The AMAC implementation options include establishing this consortium as a Federal Advisory Committee Act (FACA) body to the NSTC's Advanced Manufacturing standing subcommittee, with OSTP as the coordinating office of that NSTC AM subcommittee. AMAC should also regularly engage federal agency representatives, and be supported by the Office of Science and Technology Policy and AMNPO in this task. AMAC should be charged to:
   a. Develop roadmaps for identifying and supporting key MTAs as part of a national competitiveness strategy
   b. Develop a portfolio view of federal manufacturing investments across technologies, sectors, TRL/MRL levels and time horizons for MTAs
   c. Develop metrics to monitor portfolio progress along key performance indicators
   d. Monitor performance and recommend investment adjustments
   e. Promote best practices and sustainable collaboration models for driving adoption of advance manufacturing technologies

f. Provide feedback on public-private partnership models within the roadmap, e.g., an NNMI institute
g. Enable a framework for pipeline investments in R&D (TRL 1-4)

AMAC would approach U.S. manufacturing technology strategy development and deployment through periodic review, and with a focus on outcomes prioritized through a national perspective. Figure A1 illustrates an annual calendar of activities that AMAC could model.

Figure A1. Annual cycle of national strategy development and implementation for Advanced Manufacturing Technologies.

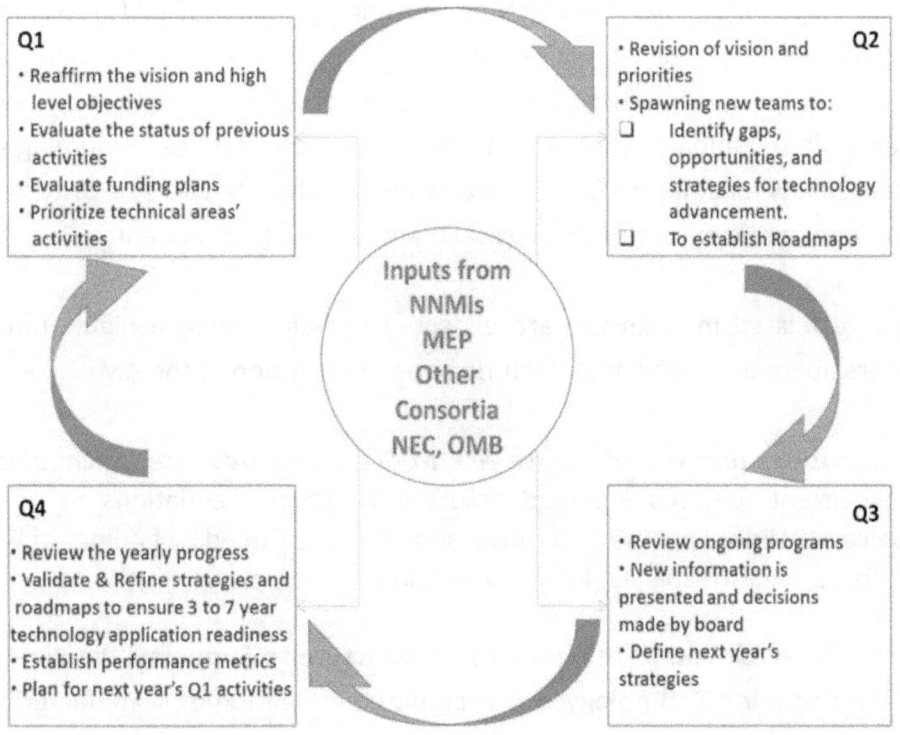

3. **Coordination of Advanced Manufacturing Research and Development:** AMP2.0 is pleased that the government has adopted AMP's recommendation of Manufacturing Innovation Institutes as part of a National Network of Manufacturing Innovation

Institutes (NNMI). However, for these institutes to be fully effective at technology readiness levels (TRL) 4-7, they will require input over time from ongoing R&D advances in relevant fields (TRL 1-3); and output to receptive industry at higher TRLs and manufacturing readiness levels. Broadly, the Advanced Manufacturing Partnership recommends establishing mechanisms to coordinate and link technology and research investments upstream from the National Network for Manufacturing Innovation to individual institutes to ensure a steady pipeline of discoveries that can be advanced to later stages of development.

In addition, to support the NNMI at both later and earlier stages of technology development, the Advanced Manufacturing Partnership identified a cross-cutting need for two new public-private research and technology efforts to spur the further development and adoption of these emerging technologies: the need for additional research and development infrastructure in the form of manufacturing centers of excellence (MCE) and manufacturing technology testbeds (MTT) to create a pipeline of earlier-stage technologies that can feed into the National Network for Manufacturing Innovation Institutes; the importance of an effort to develop missing technology standards that can de-risk the adoption of these emerging technologies, particularly for smaller manufacturers; and the importance of security at the interface between cyber systems and physical manufacturing equipment.

These three interrelated mechanisms are expected to help ensure a highly effective public-private partnership for advanced manufacturing upon conclusion of the AMP2.0 effort.

B. ***Enhance capability access and reduce risk to accelerate advanced technology adoption:*** Several persistent features emerged across MTA recommendations to overcome both technological and implementation hurdles, and were reinforced by findings of Working Team 4 (Appendix 4). We thus suggest for consideration:

   1. **New Research and Development Infrastructure to Support a Pipeline of Advanced Manufacturing Technology:** For example, the creation of Manufacturing Centers of Excellence to invest in basic research "upstream" in the innovation pipeline can advance critical discoveries important to technology areas, such as the creation of a Manufacturing Center of Excellence to address joining technologies for advanced materials, specifically advanced composites. Further, Manufacturing Technology Testbeds (MTTs) for industry to understand, customize, and test potential technologies will support SMMs. Additionally, the intermediary solutions (enhanced

MEPs), tax credits, and capital access options analyzed by Working Team 4 (Appendix 4) accelerate industry adoption of available technologies.

2. **Standards and Interoperability of Hardware and Software Systems:** Standards allow a more dynamic and competitive marketplace, without hampering the opportunity to differentiate. Development of standards reduces the risks for enterprises developing solutions and for those implementing them, accelerating adoption of new manufactured products and manufacturing methods. Industry expressed wide support of data standards for products, processes, and materials, including partnership to establish databases that were narrowly developed around specific manufactured components (e.g., an automobile side panel). This perceived advantage of standardized practices was reflected in the specific recommendations of the three MTAs we analyzed, including a subset of related findings in Table A4.

3. **Security, Data Privacy and Trust across the Supply-Chain:** Increasingly, manufacturing will proceed with automated communication among computers. To enable digital manufacturing, policies that incentivize technologies, and practices (infrastructure) that offer protection from the Freedom of Information Act (FOIA) disclosures of threat information shared with the government and/or across the supply chain, will play a key role. A focused program to mitigate security risks at the interface of the cyber systems and physical equipment in the manufacturing ecosystem that is similar to the counterfeit protection program will improve trust across the supply chain.

Suggestions for Specific Manufacturing Technology Areas for Sustainable Progress in U.S. Manufacturing Innovation

Annexes 1-10 includes ten detailed reports describing technology and implementation gaps and AMP2.0 recommendations for the three Manufacturing Technology Areas that AMP2.0 prioritized for detailed analysis and strategy. These include justified recommendations of targeted federal investments including but not limited to new institutes, as well as public-private partnerships in pre-competitive R&D and in training graduate-level experts. To the extent possible, recommendations including federal leadership or partnership include suggested agencies to implement that change. Top-level recommendations are noted in the table below.

Table A4. AMP2.0 technology strategy suggestions for three prioritized Manufacturing Technology Areas.

Technology areas:	Advanced Sensing, Control, and Platforms for Manufacturing	Visualization, Informatics and Digital Manufacturing	Advanced Materials Manufacturing
R&D Infrastructure to Support the Innovation Pipeline	▪ Establish Manufacturing Technology Testbeds (MTTs) to demonstrate the use of and business case for new technologies, including "smart manufacturing" capabilities.	▪ Create a Manufacturing Center of Excellence, focused on basic research at earlier technology development levels, on the Digital Thread.	▪ Launch Materials Manufacturing Centers of Excellence to support R&D in topics that support NNMI Institutes and other MTAs in the national strategy.
The National Network for Manufacturing Innovation	▪ Establish an Institute focused on ASCPM for energy use optimization in energy-intensive and digital information-intensive manufacturing.	▪ Launch a Big Data Institute focused on secure analysis of and decision-making via large, integrated data sets for manufacturing processes (in addition to the current Digital Manufacturing and Design Innovation Institute).	▪ Leverage supply chain management of defense assets to spur innovation and RD&D in critical materials reprocessing.

Public-Private Technology Standards	▪ Develop new industry standards, including data interoperability standards for key systems and vendor support.	▪ Craft and deploy policy standards for manufacturing cyber-physical security.	▪ Design data standards for material characterization to enable rapid uptake of new materials and manufacturing methods
Additional Strategies			▪ Establish Manufacturing Innovation Fellowships for Ph.D.s in key AMM areas, such as biotherapeutic manufacturing.

Annexes 1-10 include detailed analysis of the current landscape, vision, technical and implementation gaps, and detailed analysis of options to close those gaps to accelerate U.S. innovation and adoption of that MTA. These options were intended to achieve rapid impact over short time horizons (within 3 years) as well as sustained impact (over 10-20 years). A brief summary of these analyses for the three MTAs is provided below, noting input from a wide range of stakeholders beyond AMP2.0 members.

Advanced Sensing, Controls, and Platforms for Manufacturing (ASCPM): A new generation of networked based information technologies, data analytics and predictive modeling is providing unprecedented capabilities as well as access to previously unimagined potential uses of data and information not only in the advancement of new physical technologies, materials and products but also the advancement of new, radically better ways of doing manufacturing. Key drivers of ASCPM as a high priority MTA is that this group of related technologies is the chief technical element needed for seamless interoperation of cyber and physical assets. Our vision is that such integration via ASCPM will increase productivity, product and process agility, environmental sustainability, energy and raw material usage, and safety performance as well as economic performance—and thereby comprehensively improve the competitiveness of U.S. factories of varied sizes and complexity. In particular, broader application of ASCPM technologies has great potential in energy-intensive manufacturing, and integral to use of big data analytics to drive manufacturing decisions.

We interpret ASCPM to encompass machine-to-plant-to-enterprise-to-supply-chain aspects of sensing, instrumentation, monitoring, control, and optimization as well as hardware and software platforms for industrial automation. Although significant success has been achieved in manufacturing implementations with ASCPM, for a variety of reasons, the U.S. manufacturing industry has not come close to realizing the full potential of these technologies. These reasons include technical shortcomings of the state of the art as well as nontechnical barriers. Implementation risks include risk and initial cost; rapid rates of change in hardware and software; and limited workforce availability due to the complex and interdisciplinary nature of these technologies. Technical gaps include open standards and interoperability of devices, systems, and services; lack of real-time measurement, monitoring, and optimization solutions for machine energy consumption and waste streams; lack of low-power, resilient wireless sensors and sensor networks; and need for modeling and simulation at manufacturing-relevant scales. See Annex 1. Technical gaps were further delineated as (i) sensing and measurement gaps; (ii) control and optimization gaps; (iii) platform and framework gaps. See Annex 2.

We suggest Manufacturing Technology Testbeds as a new means for especially SMEs and SMMs to share, access, and optimize ASCPM technologies specific to narrow industry sectors; this approach will de-risk technology adoption and provide technical knowhow and access to an ASCPM skilled workforce that may be lacking within SMMs. We further suggest consideration of a Manufacturing Innovation Institute that promotes ASCMP development, specifically toward improved energy consumption monitoring and integrated smart or digital manufacturing.

Visualization, Informatics, & Digital Manufacturing (VIDM): VIDM is a set of integrated, cross-cutting enterprise-level smart-manufacturing approaches, leveraging the current advances in information technology systems and tools that will improve U.S. manufacturing competitiveness through end-to-end supply-chain efficiency, unprecedented flexibility, and optimized energy management to achieve error-free manufacturing of customized products and components from digital designs, when needed and where needed. The key drivers of VIDM are: increased R&D and manufacturing integration with end to end speed and productivity, supply chain efficiency, process yields, energy efficiency, improved sustainability; and improved process safety, flexibility, agility, configurability, and increased job satisfaction and pride. Major participants defining the VIDM landscape are in the Aerospace, Automotive and the process industry with strong support from the IT industry. Our vision is that VIDM – which includes but is broader than than the challenges expected to be addressed by the Digital Manufacturing and Design Institute (an MII) – will rapidly change the way manufacturers use and exchange information to plan, support, source, deliver, and make commercial products in the U.S. See Annex 3.

We identified and analyzed technical gaps within three distinct areas of VIDM: (i) Digital Thread; (ii) Integrated Information Systems; and (iii) Manufacturing Big Data and Analytics. (Annexes 4-6). Our findings of lowest implementation barrier and highest anticipated impact included an accredited digital manufacturing education program, including associate, undergraduate and graduate degrees; institutionalized use of digital thread tools and techniques via supplier contracts; regional MTTs and service bureaus that would help build the equipment supply base for additive manufacturing – an associated MTA that supports many early-TRL/MRL aspects of digital manufacturing; and creation of manufacturing data ontology (for standards) as well as consideration of a Manufacturing Big Data MII to support the use of voluminous, complex, and veracity-tested data across the supply chain. The analysis within these annexes may be of use to shape the focus of the DMDI, as well as supporting public-private investments such as MCEs and MTTs.

Advanced Materials Manufacturing (AMM): AMM is defined as advanced methods to design and produce a material of predictable and important functional properties within commercial products (regardless of whether that material is considered an "advanced material" such as a novel quantum dot, a "biomaterial" such as a therapeutic protein, or a "mature material" such as steel). The capacity to manufacture materials domestically, from the raw state to the fully processed or shaped material, is important for strategic interests and for the pace of innovation, including recent examples such as polymer production critical to space exploration and specialized metal production key to energy conversion devices. Key drivers for U.S. strength in materials manufacturing derive from the cross-cutting impact of materials innovation across multiple industries and sectors; historical national security and competitiveness implications of material supply uncertainties; and the potential to reduce energy consumption and environmental impact through improved materials manufacturing processes.

Historic U.S. strengths in materials synthesis and processing vary by geographic region, linked typically to abundance of the required natural resources, or to the historic co-location of the material end-user (e.g., automobiles, paper goods, airplanes). Materials design, synthesis, and processing have all advanced rapidly over the past decade, owing to both new computational predictive capabilities and new high-throughput fabrication and characterization methods. For several reasons, high-volume processing of commodity materials has shifted to other nations. Qualification methodologies, including reliability testing of such materials, is not a U.S. educational focus; the lack of such shared data slows industry adoption of new materials and methods for safety-critical products in all sectors. Finally, current U.S. strengths in reprocessing of materials at the end-of-life stage of manufactured product lifecycle exist but are segmented by material or industry sector, and focused innovation in this area lags that of other nations. Our vision is that the U.S. will train a workforce that can invent, adapt, maintain, and recycle materials critical to U.S. infrastructure, defense, medical care, and quality-of-life. This vision will also accelerate the transition from lower to higher TRL/MRL maturity, to enable faster and broader industry adoption. See Annex 7.

We identified technical gaps across all types of AMM, grouped broadly as those in (i) standardization and qualification of material properties; and (ii) labor- and resource-intensive materials synthesis and processing. We discussed opportunities for MCEs, an AMAC, and focused standards database development via existing MIIs and potential new MIIs to address these broader opportunities.

As many AMM challenges are specific to the type of material that is manufactured (metals, ceramics, polymers, biologicals, composites), we conducted three "deep dives" into specific AMM technology areas that were considered high national priority as well as diverse in technical scope: (1) Advanced Structural Composite Manufacturing; (2) Biomanufacturing of biological therapeutics; and (3) Critical Materials Reprocessing. The drivers, landscape, vision, technical gaps and analysis for each are provided in Annexes 8-10. Common points informed the aforementioned recommendations for MCEs and MTTs (Recommendation 3). Briefly, structural composites manufacturing is crosscutting and has potential to be broadly geographically distributed, but has several technical challenges that vary among industry sectors (e.g., auto vs. aerospace vs. sporting goods), so standards and MCEs focused on joining and recycling are discussed. Biomanufacturing is itself a broad topic that is delineated in Annex 9, and for biological therapeutics (antibodies, vaccines, etc.) includes technical challenges in rapid scale-up/scale-down and a depleted graduate workforce; MCEs on pre-competitive scale-up technology and public-private partnerships in graduate fellowships were proposed with broad industry interest and support. Finally, critical materials reprocessing addresses a national security and innovation concern, related to the unstable supply or price of materials that enable key technologies ranging from those in strategic defense assets to consumer electronics. Recovery, repurposing, and/or recycling of these materials can be technically and logistically challenging, but provides key advantages to supply/demand interactions for U.S. SMMs and reduce reliance on other nations within complex supply chains; MCEs and MTTs that complement existing federal investment in critical materials are proposed in Annex 10.

Appendix 2: Demand-Driven Workforce Development and Training

Background and Scope

The goal of the Advanced Manufacturing Partnership 2.0 Workforce Development Working Team was to build on the outcomes of the Education and Workforce Workstream in AMP and determine best-in-class demand-driven workforce solutions to develop technical skills and implement models that generate long term employment opportunities. The outcomes are pragmatic, scalable and sustainable. This appendix highlights some recommendations to accelerate the deployment and adoption of the most effective ways in which to develop and support the workforce of the future. Four subteams were formed to develop recommendations building upon the original AMP report findings.

Process and Deliverables

The goals of the Working Team were achieved through the efforts of four subteams called "work creeks." The team adopted a holistic approach to build scalable and sustainable solutions to address the talent pipeline challenge beginning with childhood education and ending with long-term employment models. Outcomes are in the form of Guide, Playbook and Recommendations based on current best practices. The deliverables listed below are provided as Annexes 11-24.

Work Creek 1: Advanced Manufacturing Education - Increase career pathways and "dual credit" opportunities across education (K-12 schools, community colleges, and Universities) to increase the number of qualified technical employees in advanced manufacturing.

   Deliverables:

   - Identified Exemplary Programs that engage Advanced Manufacturing and Product Realization (Design, Manufacturing, Operations, System Support) starting from the 7-12th grade level to Certificates, Diplomas, and Associate Degrees at Community Colleges, and Bachelors' and Master's Degrees at Universities.
   - Identified examples of exemplary Career Education Pathways from other industries (e.g. automobile) in order to build our own pathways model for Advanced Manufacturing.
   - Recommendations on how to duplicate scale-up and improve on the best practice programs.

- Developed a template for "pathways" model for Advanced Manufacturing training and education with multiple on and off ramps and multiple completion options (certificate, diplomas, degrees) that are stackable.

Work Creek 2: Portable and Stackable Credentialing System - Increase employer-driven, nationally portable, stackable credentialing systems through certifications and work-based learning elements.

Deliverables:

- Build awareness of credentialing through Employers, Educators and Certification of Train-the-Trainer Toolkits.
- Demonstrate evidence of success of credentialing via employer case studies and best practices at educational institutions.
- National roadmap for action with guidelines for implementation on state level, including way to address the challenges in implementing credentialing systems.

Work Creek 3: Regional Apprenticeship Models - Establish regional work/project-based study apprenticeship models with a coalition of companies and in partnership with labor market intermediaries and capture the learning and best practices from these and past Apprenticeship programs in a "How To" Instruction Manual targeted specifically for Employers the Playbook will prove the building of the apprenticeship model to be a reliable, valid and repeatable process that others can implement.

Deliverables:

- Developed an employer playbook, as a practical "How-To" manual designed primarily for employers seeking to implement an apprenticeship model as a way to build the employer's advanced manufacturing workforce pipeline. This model of apprenticeships entails completion of a community college curriculum and Department of Labor Certification, and is anticipated for publication in September 2014.
- Leveraged over 15 operating sites across all three companies to develop the playbook and initiated proof-of-concept apprenticeship program pilots across all three companies.

Work Creek 4: Bridging Modules for Veterans - Develop practical competency based "bridging "bridging modules" for transitioning Veterans focused on private sector manufacturing skills certifications.

Deliverables:

- Completed inventory of Veteran resources – key tools and applicable links that catalog and summarize the resources for Veteran use.
- Developed a practical guide for Veterans, Employers and Academic Institutions on transitioning Veterans to non-military advanced manufacturing careers.
- Gathered information on skills translators & Veterans skill badging that enables
    - Translating military experience to civilian jobs
    - Transferring competencies and skills in an electronic format
    - Alignment with current work being done on badging systems by federal agencies and other entities

The four Work Creeks worked to achieve their independent deliverables, and came together to leverage synergies where possible, and collectively achieve Workforce Working Team goals.

Appendix 3: National Network for Manufacturing Innovation (NNMI) Analysis

Background

In order to develop the steps to operationalize the activities outlined by AMP, the NNMI Working Group has engaged stakeholders on all levels in informal settings and in more formal dialog. These proposed actions include and consider these suggestions, as well as reinforce many of the activities currently being undertaken by the Advanced Manufacturing National Program Office (AMNPO).

Scope of Work

We have conducted one-on-one discussions, group discussions, have engaged stakeholders at AMP2.0 Regional Meetings, have reached out to existing Institutes as well as other AMP2.0 Working Groups and have invited external participation in our deliberations. We have also considered the input of experts who have generously offered their insights as well as the previous work product from within and outside of the Administration.

Key Findings

In order to help ensure existing and near term Institutes have what they need to be successful, we have focused on fundamental concepts and recommendations that will position the NNMI to realize its full potential in supporting/growing U.S. manufacturing, innovation and the supply chain, ultimately growing regional economies and creating jobs. We have captured our findings in six Letter Reports that address the following areas, which we believe are keys to the NNMI's success (Annex 25):

1. Internal and External Communications
2. The NNMI Narrative
3. Network Governance & Operations – Key Considerations
4. Network Governance & Operations – Organization & Structure
5. Intellectual Property
6. Technology Area Identification

Recommendations

Annexes 25-31 include detailed support for the findings on the following six topics:

1. ***Internal & External Communications***: The focus is to provide guidance on how to better ensure individual Institute effectiveness and operations as well as overall Network effectiveness by increasing the stakeholder's ability to leverage the network model. Recommendations include mechanisms to help to provide consistent messaging on the value of the NNMI, a communications plan and assistance with both internal and external communication processes.

2. ***Narrative***: The Narrative is intended to provide context for the NNMI and to provide an organized flow of main talking points to help communicate the value proposition. The Narrative is a general messaging guideline to better ensure consistent messaging and clarity with regard to the value of the NNMI to stakeholders and ultimately build participation and support.

3. ***Governance & Network Operations – Key Considerations***: In developing the governance and operational structure of the Network, there were several important considerations that warranted attention. Prime considerations included:
   - The Benefits of the Network – the value of the Network to the Institutes and external stakeholders
   - Diversity – ensuring that all perspectives are represented so as to ensure the needs of all stakeholders are met as intended
   - Balancing Network Consistency with Institute Autonomy – Insuring the Institutes can function while allowing for reasonable, productive consistency
   - What the Network is…and Isn't – helping to clarify the roles of the Network and set expectations
   - Workforce Development – ensuring that the NNMI is leveraged to better ensure education is demand driven
   - Metrics – providing general guidance on how to develop proper metrics
   - Leveraging the Manufacturing Extension Partnership – helping to ensure that the MEP becomes a major "tool in the NNMI toolbox"
   - Network Expansion – Provide general guidance and suggestions on how to promote greater collaboration

4. ***Governance & Network Operations – Organization & Structure:*** A key to ensuring that the NNMI (Network) remains connected to those it is intended to serve and meets its intended purpose is to institutionalize a structure that includes representatives from all stakeholders in its governance and in various advisory groups. The proposed structure provides guidance on how to strike a balance between industry, government and academia and ensure all perspectives "have a seat at the table".

5. ***Intellectual Property Management:*** In addition to providing a summary of key best practices, the Letter Report on this topic provides general guidance on how to best manage IP so as to promote collaboration and best facilitate innovation within the NNMI setting. Findings address issues related to areas such as Background IP, Data Rights, Publications, Government Rights and Revenue Models.

6. ***Technology Area Identification:*** Rather than debate various technology areas that may make sense to pursue at this time, the approach outlined in the Letter Report on this topic is to provide guidance on how to choose technology areas to pursue via the NNMI at any time in the future. Included in the said report are a series of questions intended to help identify where investments should be made via Institute technology focus areas. Key considerations in developing these questions include ensuring that:
   - The process is outcome driven
   - The investment will engage a diverse landscape of stakeholders
   - There is reasonable evidence that the proposal has the potential to generate economic value (the technology has a path to commercialization and will attract investment), advances national security and helps to sustain competitiveness
   - The required supply chain either exists or can be developed (includes considerations on how to grow the required supply chain)
   - That the necessary workforce exists or can be developed.

Conclusions

The above proposals on communication, governance, IP management, and MII topic selection are provided to promote the success and longevity of the NNMI. Written guidelines that afford clear communication of these policies, for both the current and potential funding scenarios, will be important components of that success.

Appendix 4: Scale-Up Policy

Background

As part of this effort, AMP 2.0 initiated a work stream on "Scale-Up Policy" with the charge to make recommendations about improvements to the business climate that will foster development of small- and mid-sized manufacturers, both start-ups and established enterprises. We have identified activities that will foster the flow of information, capital and expertise to these critical members of the manufacturing ecosystem, increasing their likelihood of success as they implement new manufacturing technologies.

Scope of Work

The Advanced Manufacturing Partnership Steering Committee Scale-Up Policy Group (Work Stream 4) was created in October 2013. The committee draws from the AMP 2.0 member organizations from labor, academia and industry. The team has met through biweekly conference call and in several face-to-face meetings concurrent with the AMP 2.0 regional meetings in Atlanta, Akron, Albany, Boston, Detroit, and Washington DC. The team identified separate supply chain issues and capital access, investigating each individually while communicating findings across the entire team. The expertise of the participants was buttressed by literature research and supplemented by sessions with experts and listening sessions at the regional meetings. The input and perspectives of many financial experts, U.S. manufacturers, and economic development organizations through regional breakout discussions, national roundtables and one-on-one meetings were invaluable in developing the recommendations. Major stakeholder and expert forums are documented in the Supplemental Information.

Key Findings

Scale-up can be defined as the translation of an innovation into a market. There are significant technical and market risks faced by new manufacturing technologies during scale-up. The path to successful commercialization requires that technologies function well at large scale and that markets develop to accept products produced at scale. It is a time when supply chains must be developed, demand created and capital deployed. To compete globally and be a leader in innovation, the United States needs to significantly improve its ability translate innovation into practical production. To achieve commercial scale with promising advanced manufacturing technologies requires three things: 1) networked supply chains, 2) the rapid diffusion of technology through the networked supply chains, 3) access to capital. Barriers to achieving scale scale are **impeded flows** of technical / market insights, supply network relationships, and funding.

- **Flow of Technical and Market Insight:** Most small and medium sized manufacturers (SMM) lack information about emergent business opportunities enabled by advanced technology and approaches to enter those markets profitably. SMM efforts are often fragmented and siloed.
  There is a need to provide market insight to SMMs including business opportunities, technology readiness and entry strategy. Market Insight can increase the number of SMMs considering technologies, accelerate and de-risk adoption.
- **Flow of Relationships (supply network deals/development):** When pursuing opportunities, new and established small and medium sized manufacturers have difficulty finding, developing and managing relationships with supply chain partners and capital sources needed for scale-up. There is a need to identify ways to better share, develop and transfer knowledge across the supply network regionally and nationally and reduce hurdles for new supplier/vendor agreements.
- **Flow of Capital:** Manufacturing frequently does not offer the short-term returns necessary to attract capital. Returns are limited by the capital required to build a manufacturing facility, the time required to garner significant revenues and the risk of deployment of unproven technology. Investment can be made more attractive by reducing capital at risk, speeding up time to revenue and by providing expertise to reduce chance of failure. The team focused on incentives to increase private-sector investment, recognizing that this is a way that manufacturing can most help itself. Demand creation, sharing of expertise and investment from established players all reduce risk associated with manufacturing investments.

The United States can overcome these barriers and catalyze the flow of technology-driven manufacturing innovation with solutions that broker the transfer of knowledge nationwide, increase technical and market insight available to SMMs, and open up capital access by reducing financial risk and creating greater awareness of solutions to SMMs. Currently, support for manufacturing is largely regional and varies substantially across regions.

Proposals

Detailed in this letter is supporting information for policy recommendations that we deem central to Advanced Manufacturing Scale Up strategy and policy. Analysis was organized in the following areas:

- **Relationships**: Broker exchange of information nationwide through Enhanced and Expanded Intermediary Solutions

- **Technical & Market Insight**: Deliver Market and Technical Insight to the capable SMM supply network at earlier TRL/MRL stage
- **Capital Access**: Open up capital access by reducing financial risk, creating and providing greater awareness of solutions to both Established SMMs and Start-Up Firms

Specific to Relationships: Brokering the exchange of information nationwide through Enhanced and Expanded Intermediary Solutions

Information asymmetries are resolved by effective intermediaries. Intermediary entities direct and connect small and medium sized manufacturers to the diverse resources for technology commercialization and scale up including: 1) subject matter experts, 2) relevant technology, 3) knowledge assets, and 4) capital sources.

Successful supply chain development and technology diffusion programs and institutions that run them currently vary geographically, and by industry or technology. In addition, the small and medium sized manufacturer's space represents a dynamic group of firms rather than a fixed set. As such, the intent of these analyses is to focus on *the required characteristics* of enhanced intermediary solutions rather than to prescribe a formal model. Intermediaries may have different functions in different places or in different industries and must successfully meet the needs of the firms and industries they serve. There are existing intermediaries, such as the Department of Commerce's Manufacturing Extension Partnership (MEP) and the National Network for Manufacturing Institutes (NNMI). There are also industry groups, such as the American Chemical Society's Entrepreneurial Resources Center. The goal moving forward is to enhance and improve the intermediary services provided by current institutions (MEP Centers and NNMI as examples) and new entrants, in order to more productively address the challenges faced by SMMs in adoption of new technologies. MEP and NNMI have fulfilled some of the services effectively, however a market for providers should lead to new and improved productivity of existing intermediary services. Note that NNMI tends function at the *Technology Readiness Levels* of 4-7 whereas there is a significant need for scale-up support in *Technology Readiness Levels* 7-9.

We found that several types of organizations currently provide firms with some of the key characteristics we itemize below (see Design Characteristics of Intermediaries). These organizations include private sector, public sector, and non-profit intermediaries. Illustrative examples are included in the Appendix A. What is consistent among the successful examples is that: 1) they are regional in scope ("regional" meaning a coherent economic ecosystem not an administrative designation), 2) their focus is technology or industry specific, and 3) they provide information, technical assistance, and other services to a network of firms rather than a single firm. The Manufacturing Extension Partnership is one example of a public sector intermediary that meets some of the Design Characteristics itemized below. However, many other examples exist including NorTech (OH), the Rochester Regional Photonics Cluster (NY), The Great Lakes Wind Network, SF Made (CA), Maker's Row (NY/NJ), and Life Science Alley (MN).

Outlined in the list are key design characteristics critical for success. These characteristics provide a recipe that can be baked into different institutional forms to meet needs and utilize assets from the differentiated landscape of U.S. manufacturing, academia, and local/state/federal economic development agencies.

<u>Design Characteristics and Requirements</u> (for affordable, accessible, and state-of-the-art intermediary services)

a. Effective identification & understanding of challenges and opportunities including those related to workforce, technology, capital, business environment, and other components of the manufacturing ecosystem.
b. Technology maturation resources accessible to manufacturers in terms of cost and utility (for example: "microlabs" providing a hands-on capability, testing, prototyping, pilot processes, technical assistance)
c. Market insight and related information on the potential demand for emerging technologies in order to reduce risk and enhance scale-up opportunity.
d. Certification of suppliers and technologies (testing, licensing, certification, & supply-chain matching)
e. Asset mapping of regional resources including research and development and workforce development resources
f. Connections to industry specific supply chains in other regions and as well as in global networks (expanding supply-chains across scales)
g. Real-time and up-to-date knowledge sharing about firms in the local/regional supply chain to aid in succession-planning, matching, and technology diffusion
h. Understanding of viable financing sources and relationships to facilitate referrals
i. Ability to efficiently allocate intellectual property rights to optimize technology development and commercialization
j. Effective monitoring and advocacy as appropriate for optimal governmental regulation and industry standards that support adoption of new technologies and corresponding scale-up.

k. Effective reduction and consolidation of steps in the manufacturing processes driven by advances in knowledge and technologies including improved manufacturing processes, materials, and sensors among others.
l. Effective marketing and distribution of relevant enhanced training services for all persons in the manufacturing design and implementation process.

We suggest an **open request for proposals** for new and existing intermediaries to deliver the services outlined in the Design Characteristics/Requirements section above with complementary increased visibility of such intermediaries. This reinforces the implementation of Recommendation 3 in PCAST's 2012 AMP report which stated: "Establish a National Network of Manufacturing Innovation Institutes: Manufacturing Innovation Institutes (MIIs) should be formed as public private partnerships to foster regional ecosystems in advanced manufacturing technologies. These MIIs are one vehicle to integrate many recommendations" (p. 12). We propose that providers of our recommended "enhanced intermediary solutions" serve as the vehicles facilitating the adoption of advanced technologies by small and medium sized manufacturing enterprises. The increased visibility should include direct and targeted communications and an enhanced web site presence on manufacturing.gov. Where appropriate, existing intermediaries, including MEPs and NNMIs, could be reconstituted consistent with the above design characteristics. Appropriately redesigned next generation MEPs have a particular opportunity to assist SMMs with scale-up. In some regions they could serve as the regional coordinator or as a one-stop shop for enhanced intermediary services. The NNMIs may be in a position to hand off projects (particularly in the TRL 8-10 range) to providers of the enhanced intermediary solutions, such as the next generation MEPs.

Specific to Technical & Market Insight: Delivering Market and Technical Insight to the capable SMM supply network at earlier TRL/MRL stage

Small and medium sized manufacturers represent 89% of firms in the U.S. manufacturing supply network and 46% of manufacturing employment[2]. SMM make investment choices based on an understanding of business opportunity as do their larger peers. However, small firms often lack the internal capacity to analyze potential opportunities due to insufficient information about the potential demand generated by new technologies and understand how to enter those highly dynamic markets profitably. Thus, technology adoption by the existing supply network is limited and delayed. Further, financing for small firm scale-up is dependent on a compelling case

---
[2] Statistics of U.S. Businesses (SUSB), www.census.gov

requiring information about market demand and a clear path to commercialization. For these reasons, affordable and accurate market and technical insight is critical for SMMs to develop an entry strategy, and mobilize resources to adopt new technologies for processes, materials, and new products. The existing NIST MEP program has the ability to provide SMMs with market insight through fee for service projects but is capacity constrained and service models are one-to-one projects covering the full spectrum of potential technologies (e.g., MEP-led Technology-Driven Market Intelligence (TDMI) program). Furthermore, the recently launched MIIs could provide a mechanism to deploy market and technical insight across a capable SMM network, and at an earlier TRL/MRL stage to encourage SMM investment in capability development in parallel and in collaboration with technology development.

SMM entrants, collectively, must either possess or develop the necessary capabilities to compete in the future supply network in order to successfully scale up a technology. Key elements of the suggestions are described below and based on a pilot that examined a specific Technology (Printed Electronics), Region (Midwest), and Established SMM (Mac Arthur Corporation). The pilot was sponsored by Mac Arthur Corporation, University of Michigan, the IRLEE TCA Program, and the National Center for Manufacturing Sciences between May and August 2014.

Findings from the Printed Electronics Pilot are expected to be broadly applicable to all Advanced Manufacturing Technologies to accelerate scale-up and include the following elements.

- Create a "Go-To" Supply Network Hub: Where there is sufficient public/private demand, support the creation of a regional hub where technology and industry network expertise resides and collaboration can be promoted. Provide visibility of the hub to current and potential network participants and position as the "Go-To" resource for supply network companies and SMMs entering or participating in the market.
- Create and Maintain Platform Knowledge Assets: Utilize industry and technology experts, and robust methodologies to generate and maintain platforms of knowledge that can be utilized across the network of SMMs to rapidly scale-up the supply network. Key knowledge assets should include: Technical & Manufacturing readiness/hurdles, Supply network and national resources map, SMM market entry opportunities by industry, capital and capability requirements for entry, and a framework for individual SMMs to rapidly and cost-effectively prioritize attractive and best-fit areas of focus. Lessons learned should be aggregated and utilized from supply network development of previous advanced manufacturing technologies in the U.S. to de-risk small and medium sized manufacturer's entry.
- Attract Supply Network Participants, especially SMMs: Publicize market and technical insights via forums and workshops where relevant industry clusters and/or technology

are being accelerated (e.g., NNMI, industry clusters, regional clusters). Workshops are intended to draw the interest of strategic industry OEM partners, SMM supply network participants, technology providers, and capital sources. Workshops should also target associations of existing SMMs that have capability and capacity to benefit from the advanced manufacturing technology.

- Provide Market Entry Support: Connect interested and viable small and medium sized manufacturers to service partners that can help them develop and execute their plans. Leverage the platform of knowledge and insights for efficiency and affordability. The enhanced intermediary solutions described in detail above could serve as a mechanism to provide this support as could appropriately scoped and resourced MEP centers.

Platform tools and frameworks developed during the pilot will be redacted and shared with project sponsors and AMP2.0 for use in implementing this recommendation across advanced manufacturing technology areas and with potential service providers. Anticipated timing for the release of platform tools and frameworks is November 2014.

The Great Lakes Wind Network (GLWN.org) and The New York Battery & Energy Storage Technology Consortium (ny-best.org) are examples of creating market insight at scale and utilizing such information to create the supply network. These models provide market and technical insight leading to efficient and rapid supply network creation.

Specific to Capital Access: Increasing capital access by reducing financial risk, creating and providing visibility of solutions for established and start-up firms

Manufacturers have particular financing needs as has been documented elsewhere, such as in NIST MEP Report, *Connecting Small Manufacturers with the Capital Needed to Grow, Compete and Succeed*. There is money available in capital markets but manufacturing opportunities are frequently not compelling investments. As a result, SMMs looking for growth capital often have trouble finding it. Established and new SMMs face many of the same challenges. Advanced manufacturing SMMs often are not compelling investments due to technology risk, market adoption risk, long lead times to significant revenue and significant capital requirements. Capital will flow to SMMs when risks are reduced and returns are competitive. Reducing capital requirements is one way that government has encouraged manufacturers through grants, loan guarantees and tax deferrals. Frequently overlooked are the other means that investments can be made more attractive to capital markets through demand creation (reducing market adoption risk), providing access to technical expertise (reducing technical risk), and through reducing development time.

Established SMMs and new entrants can benefit from many of the same programs. The following additional points speak to federal and regional sources of capital:

- *Better Streamlining, Advertising and Coordination between Federal Funding Sources Geared Towards Manufacturing* – Over 30 federal programs have been identified as providing financing to manufacturers. We highlight eight that we think are most applicable to the scale-up challenge, largely from DOD, DOE and the SBA. The challenge with most of these programs is that they provide relatively small amounts of capital relative to the needs of commercial scale up. Appendix B documents known programs at this time and data sources. Grants and loan guarantees are one tool that the government has used that has proven successful in fostering advanced manufacturing. In particular, DOE and USDA grants and loan guarantees have been applied to many projects in biofuels, energy storage and renewable energy. Since the start of the Obama administration, USDA Biorefinery Assistance Program has provided approximately $684 million in assistance. The Department of Energy's (DOE) Title XVII Loan Guarantee Program and the Stimulus add-on known as the 1705 program combined for a total of 28 loan applications were finalized worth approximately $15 billion. These are examples of programs that have provided significant capital. While not specific to manufacturing, these programs are both large and clearly were successful in accelerating advanced manufacturing in this sector. Future programs should be considered part of the manufacturing effort, advertised and coordinated with other manufacturing specific efforts.

- *Increase Visibility of Current National Banks and Other Investors that are Engaging with Manufacturers and Recruit Others.* While there is overall agreement that traditional banks have pulled back from lending to SMMs, some national and regional banks have been actively engaged with manufacturers. Their strategies and success should be highlighted to encourage other banks and investors to invest more in the manufacturing sector. PNC is one example of a bank that is actively engaging with manufacturers. Information about such lenders could be tracked by an intermediary entity, a centrally visible data clearinghouse, or contained in or partnered with the National MEP network.

- *Extend the Research and Development Tax Credit* –Extend the Research and Development Tax Credit - The existing R&D tax credit should be reauthorized and potentially modified to allow for early stage domestic testing of commercial viability to be covered including investments in depreciable property (RD&D – Deployment).

The credit would be made refundable to ensure that firms that do not yet have a source of revenue or are currently operating in a loss situation would benefit immediately, when the demand for funds is most critical.

- *Public-Private Investment Fund* – Provide government loans to private-sector investors in "first-of-a kind" production facilities at either the pilot or commercial scale. Existing programs at both Treasury and the SBA provide models for this type of fund, which would support projects spread across a multiplicity of funds and a diverse set of fund managers. The fund would award loans to investment funds or investor consortia in an equivalent amount to half the cost of the project being financed. The scale of the fund would be significant in order to support investments of at least $40 million. A credit subsidy of $500 million would support $5 billion in debentures and a total of $10 billion in total investment.

- *Facilitate Connections among Corporate Strategic Partners and SMMs* – Strategic partners can play a critical role deployment of advanced manufacturing technologies. Partnering with a SMM enables the strategic to access new technologies while reducing its exposure. The SMM can gain supply chain access, market access, equity investment and/or technical expertise as a result of the partnership. Creating better visibility into SMM investment opportunities and technologies for interested corporate strategic partners can help lower search costs while providing potentially important scale-up partners to start ups. Interactions of SMMs with potential strategic partners are fundamentally asymmetric in nature and fraught with peril. While a single strategic investment can make a small company successful, any single investment is unlikely to be viewed as critical to the strategic investor. Match-making opportunities abound, but are inefficient and are not easily found by SMMs. Forums are organized by both the private and public sectors, by for-profit conference companies, trade associations, industry groups, federal agencies and regional intermediaries. SMMs are frequently inadequately prepared with appropriate business plans and investment prospectuses. The government should invest in, and be a customer of, a private-sector web-based platform that creates a portal for strategic partners to learn more about start-ups and their technologies. Local, regional and national expertise in development of actionable business plans should also be buttressed.

- *Use Tax Incentives to Foster Investment in Manufacturing and Partnerships in the Manufacturing Ecosystem* – The New Markets Tax Credit Program (NMTC) was established in 2000 to increase investments in low-income communities. They are suitable and have been used by both start-ups and established companies. The first awards were granted in 2003 and, over the last decade, 836 awards allocating a totaling over $40 billion. Envisioned as a catalyst for effective public-private partnerships, NMTCs are largely delivering on that promise. This program has both good and bad features as related to manufacturing. The investment must be in a low-low-income census tract, of which there are many. This does mean that existing businesses may find themselves are ineligible if they do not sit in an appropriate tract. The program is complex and intermediaries are required. The program is oversubscribed, with requests exceeding supply by approximately 7X. The program brings approximately $8 of private funding for every $1 of federal support. The support is in the form of a tax credit to qualified equity investors. As such, established established companies with a federal tax liability are incentivized to make investments which offset that liability. They are investors, incentivized to make the investment entity successful. To date, 23% of the projects funded, the largest single share, are manufacturing projects. Representing over $1 billion of investment, these these projects created over 5082 FTE jobs and 1611 construction jobs. The AMP effort effort has identified that this program is not well known even among professionals in intermediary agencies. Comfort with the program among participants is now well established; the incubation period is clearly over and supply is woefully inadequate. We suggest that the NMTC program continue and that additional allocations should be considered based on the success of the program in attracting capital. The program program also serves as a model for a wider manufacturing tax credit program that could soften the census tract citing requirements. Expansion of New Market Tax Credits or similar vehicles targeting manufacturing should be considered as a means for building the manufacturing ecosystem through support of new and expansion projects; potential for Manufacturing Renaissance Tax Credit.

- *Government Procurement and Demand for SMM Technologies*– Leverage existing federal programs, like ManTech in DOD, and Buy America authority (for example, in DOT) to make scale-up investments for strategic technologies and provide access to long-term demand through, for example, multi-year purchase orders that could be capitalized by manufacturers looking to scale. DOD, in particular, has experience that could be leveraged from its recent efforts on energy efficiency. Strategic technologies should include an expanded definition of those technologies deemed important to national security to include energy and water-related technologies.

There are differences between start-ups and established SMMs attempting to move into advanced manufacturing. One major difference is the ability of start-ups to attract venture capital funding. Many high-risk manufacturing initiatives have been funded by U.S. venture capital investors. In most cases, venture capital is characterized by the desire for an exit event, in the form of an IPO or acquisition. The AMP team research found that reporting rules enacted in the U.S. have adversely impacted the IPO market, impeding the ability of small companies to explore traditional equity markets for capital. Recent data indicate that U.S. IPOs have returned to more traditional levels, raising some question as to the impact of regulations.

Start-ups also differ from established SMMs in that they are rarely profitable. Federal tax deferrals provide no benefit. Local and regional incentives can reduce cost of operation and should be encouraged for all manufacturing. See Annex 26 for supplemental information and references.

Appendix 5: Manufacturing and Engagement

Background

American manufacturing is at a crossroads. The manufacturing share of the U.S. GDP has declined from 33% in the 1950s to 12.5% in 2012. During the decade leading to the "Great Recession" of 2008-2009, the manufacturing sector lost approximately 6 million jobs, from an employment high of 17.5 million in 1998 to 11.5 million at the end of 2009.[3,4] While more than 500,000 of those jobs have been regained since the deepest point of the recession, that dramatic job loss has fed perceptions that American manufacturing is declining and that manufacturing jobs are moving offshore. Recent polling demonstrates this sense by voters, with 50% of respondents believing that the U.S. is continuing to move more jobs out of the U.S. for overseas versus only 13% that believe jobs are coming back.[5] This uncertainty as to the long-term viability of American manufacturing and manufacturing careers has created a problem in attracting the next generation of workers to careers in this manufacturing sector.

On the other hand, more than 60% of voters agree that manufacturing is the single most important part of the American economy and we need a manufacturing base here if this country and our children are to thrive in the future.[5] Manufacturers in the United States are the most productive in the world, far surpassing the worker productivity of any other major manufacturing economy. Manufacturers in the U.S. perform two-thirds of all private sector research and development in the nation, driving more innovation than any other sector. This innovation and growth potential, as well as the jobs that are being created in American manufacturing, make properly developing a workforce to fill those positions a key priority. If done properly, it can create a virtuous circle by which employers are incentivized to bring even more jobs onshore and drive even more investment in America.

---

[3] M. Barker, "Manufacturing employment hard hit during the 2007–09 recession", *Monthly Labor Review*, April 2011.
[4] National Association of Manufacturers, "Facts about Manufacturing in the United States". http://www.nam.org/Statistics-And-Data/Facts-About-Manufacturing/Landing.aspx
[5] The Mellman Group and North Star Opinion Research, "Findings From a National Survey and Focus Groups of Likely 2014 Voters", poll and focus groups conducted for the Alliance for American Manufacturing (January 2014).

This growth is critical to sustaining the economic recovery and creating and retaining good jobs at every educational level for American workers. As such, corporations, academic institutions, labor organizations, and government must all work in unison to both create a new image of manufacturing that is "**interesting, innovative, impactful, and most of all *increasing*.**" Further, this work must ensure that the jobs created are durable and stable paths to the middle class, which if done correctly will enable generations of young people to once again see manufacturing careers as viable, well-compensated, and stable career paths.

Scope

The Advanced Manufacturing Partnership 2.0 (AMP2.0) was chartered in the Fall of 2013, and a "Manufacturing Image and Engagement" team was assembled and charged to:

- Consider the elements of an awareness campaign to improve image of manufacturing,
- Develop an outreach program for support,
- Host and participate in AMP2.0 Regional Meetings, and
- Evaluate the effectiveness of conducting a National Manufacturing Summit.

In this appendix, we summarize our work and provide suggestions on ways in which the image of manufacturing can be improved.

Process & Deliverables

Over the last eight months, the Manufacturing Image Team carried out the following workshop and engagement activities to define our message and approaches to the manufacturing image campaign:

- We conducted a manufacturing image round table in Washington DC. Invited participants included representatives of the Society of Manufacturing Engineers (SME), American Society of Mechanical Engineers (ASME), and the Manufacturing Institute.

- We held an all-day workshop on manufacturing image. Participants included the Image team members, SME, the Ad Council, Manufacturing Institute, NIST Manufacturing Extension Partnership, and members of the Advanced Manufacturing National Manufacturing Program Office.

- Participated in the AMP regional meetings held at the University of Akron, RPI, and MIT.

- Hosted a national meeting in Detroit in collaboration with the Big M conferences. The national meeting was organized by SME, the University of Michigan and Northrop Grumman Corporation. The Big M was organized by SME and included six collocated conferences and technology exhibits, as well as a keynote address by U.S. Secretary of

Commerce Penny Pritzker. The AMP national meeting also included a keynote by Jason Miller of the National Economic Council.

Through these focused workshops and AMP 2.0 regional and national meetings, we have solidified our recommendations around two actions of highest priority for the government to consider:

1. Develop a national manufacturing image campaign focusing on the current and potential future growth of American manufacturing to combat perceptions of the sector as declining.
2. Support National Manufacturing Day as a way to engage manufacturers with schools and communities about the benefits and impact of manufacturing in communities.

Findings

We suggest targeting students at all educational levels, primarily K-12, school counselors, parents, and the general public in the manufacturing image campaign.

1. Develop a national manufacturing image campaign focusing on the current and potential future growth of American manufacturing to combat perceptions of the sector as declining.

We propose a layered approach to the manufacturing image campaign. One-shot national campaigns, while important, will not be effective without simultaneous regional and local deployment over a variety of platforms, including social media and specific activities for each age group.

Steps that need to be addressed and followed in developing the national campaign include:

- Selection of a media agency to create and execute the image campaign.
- Alignment with industry and professional organizations to extend the reach of the message and follow up
- Coordination with academic institutions, STEM initiatives, and workforce training initiatives
- Coordination with other federal programs and departments on messaging
- Enhancing the visibility of National Manufacturing Day as a platform for manufacturers to showcase their own images that reinforces the message at the local level.

2. Leverage National Manufacturing Day as a way to engage manufacturers with schools and communities about the benefits and impact of manufacturing in communities.

An excellent approach in regional and local engagement already exists with the National Manufacturing Day. Co-produced by FMA, NAM and the NIST MEP, the National Manufacturing Day "addresses common misperceptions about manufacturing by giving manufacturers an opportunity to open their doors and show, in a coordinated effort, what manufacturing is — and what it isn't. By working together during and after MFG DAY, manufacturers will begin to address the skilled labor shortage they face, connect with future generations, take charge of the public image of manufacturing, and ensure the ongoing prosperity of the whole industry."[6]

We propose that AMP and the DOC Manufacturing Council join forces in promoting the National Manufacturing Day so that companies all over the country will open their doors to students and the public so that we can work together to create a new image for manufacturing.

Conclusions

The specter of 6 million lost manufacturing jobs, communities left bereft, and workers laid off will not be erased overnight. Rebuilding American manufacturing is a multifaceted process, but it is also one of the most critical projects to which every level of American society can and must dedicate itself.

One critical step in this process is ensuring that as employers create new manufacturing jobs there are suitable candidates to fill them. Doing so requires that potential applicants receive the training they need to fill these jobs, a task given to another team of the Advanced Manufacturing Partnership. But it also requires ensuring that the potential next generation of the American manufacturing workforce is incentivized to seek out careers in manufacturing. For decades, workers flocked to manufacturing careers because those jobs were viewed as stable, solid careers that provided a path to the middle class for workers at every educational level. That perception has been shaken by the job losses of past decades, and rebuilding it requires two things. First and foremost, it requires continued, real, and sustainable growth in these jobs. But it also requires convincing people that this job growth in manufacturing has occurred and will continue.

A comprehensive, multiplatform campaign to drive home this message that the American manufacturing sector is growing again is a key component in bringing about this attitude shift. We urge its adoption.

---

[6] http://www.mfgday.com/

Appendix 6: Abbreviation Glossary

ABET	Accreditation Board for Engineering and Technology
AMNPO	Advanced Manufacturing National Program Office
AMP	Advanced Manufacturing Partnership
AMP2.0	Advanced Manufacturing Partnership 2.0 (successor to AMP 2011-2012 effort)
AMP2.0 SC	Advanced Manufacturing Partnership 2.0 Steering Committee
DARPA	Defense Advanced Research Projects Agency
DOC	Department of Commerce
DOD	Department of Defense
DOE	Department of Energy
GDP	Gross Domestic Product
IIE	Institute of Industrial Engineers
IP	Intellectual Property
ITC	Investment Tax Credit
ITIF	Information Technology & Innovation Foundation
M-TAC	Manufacturing Technology Acceleration Center
MAPI	Manufacturers Alliance for Productivity and Innovation
MCE	Manufacturing Center of Excellence
MEP	Manufacturing Extension Partnership
MTT	Manufacturing Technology Testbed
NAM	National Association of Manufacturers

NDEMC	National Digital Engineering and Manufacturing Consortium
NEC	National Economic Council
NIH	National Institutes of Health
NIST	National Institute of Standards and Technology
NNMI	National Network for Manufacturing Innovation
NSF	National Science Foundation
PCAST	President's Council of Advisors on Science and Technology
R&D	Research and Development
RD&D	Research, Development & Deployment
SBIR	Small Business Innovation Research
SME	Society of Manufacturing Engineers
SME's	Small and Medium-Sized Enterprises
SMMs	Small and Medium-Sized Manufacturers
STEM	Science, Technology, Engineering and Mathematics
TAACCCT	Trade Adjustment Assistance Community College and Career Training
TRL/MRL	Technology Readiness Level / Manufacturing Readiness Level
U.S.	United States
WTO	World Trade Organization

www.ingramcontent.com/pod-product-compliance
Lightning Source LLC
Chambersburg PA
CBHW080707190526
45169CB00006B/2273